Interviewers' Deviations in Surveys

SCHRIFTEN ZUR
EMPIRISCHEN WIRTSCHAFTSFORSCHUNG
Herausgegeben von Peter M. Schulze und Peter Winker

Band 22

PL ACADEMIC RESEARCH

Peter Winker / Natalja Menold / Rolf Porst
(eds.)

Interviewers' Deviations in Surveys

Impact, Reasons, Detection and Prevention

PL ACADEMIC RESEARCH

Bibliographic Information published by the Deutsche Nationalbibliothek
The Deutsche Nationalbibliothek lists this publication in the Deutsche Nationalbibliografie; detailed bibliographic data is available in the internet at http://dnb.d-nb.de.

Cover Design:
© Olaf Gloeckler, Atelier Platen, Friedberg

ISSN 1437-0697
ISBN 978-3-631-63715-9

www.peterlang.de

Interviewers' Deviations in Surveys – Impact, Reasons, Detection and Prevention

Peter Winker, Natalja Menold and Rolf Porst

Topic

Survey data still represent a major, if not the major input to empirical research in all fields of social sciences ranging from sociology and political sciences to economics, but also in neighboring fields of research, e.g. in psychology or educational research. While other modes of data collection, in particular online surveys, gained importance over the last couple of years, interviews conducted by interviewers still play a prominent role when it comes to collect high quality data. This applies particularly to face to face interviews.

There exists a huge and still growing literature on the effects of survey mode on data quality addressing issues related both to sampling and non-sampling errors. Also framing effects resulting from the form, structure and specific wording of a questionnaire have been analyzed in some detail. Finally, interviewer effects in the sense of affecting respondents' behavior have been subject to analysis. All these components are highly relevant for the generation of high quality data. However, a further aspect appears to be considered in less detail, at least when judging according to published research, namely interviewers' deviations from prescribed standards in surveys. Obviously, such deviations might affect data quality at least to the same extent as the other topics just mentioned.

Given the relative scarcity of literature on interviewers' deviant behavior, a workshop was held at the Justus-Liebig-University aiming to collect international contributions to the field and fostering scientific exchange. The workshop on "Interviewers' Deviant Behaviour – Reasons, Detection, Prevention" was organized in context of a project funded by the German Research Foundation (DFG; WI 2024/4-1, ME 3538/1-1) by Christoph J. Kemper, Natalja Menold, Nina Storfinger and Peter Winker in Castle Rauischholzhausen, October 27 – 28, 2011.

This volume collects peer reviewed papers based on the presentations given at that workshop. It is organized in three sections addressing specific issues discussed: methods for identification of deviant behavior, detection and prevention of falsifications in survey practice, and interviewers' behavior and its effect in survey data. Thereby, this volume provides a rather unique collection both of case studies illustrating deviant behavior, which seems to occur much more frequently than reported in the literature, and of methods aiming at identifying and preventing such misconduct. We will shortly present the individual contributions in the next section of this introduction before turning to an outlook to further topics, which deserve more research input, and to possible implications of the collected findings in this volume with regard to survey methodology and empirical research in general.

Content

The first part of the volume is devoted to data based method for identification of interviewers and interviews which might be considered "at risk", i.e. which show traits typical for deviant behavior.

The literature review by Bredl, Storfinger and Menold indicates that the issue of falsifications was discussed for a long time, but only a quite limited number of publications devoted to the issue can be found. Nevertheless, their survey supports the assumption that falsifications in surveys are a relevant phenomenon, that falsifications might substantially affect the qualitative outcome of survey based research, and that there exist properties of falsified data which might be used to identify results of deviant behavior.

Blasius and Thiessen suggest identifying poorly conducted interviews based on repeated answering patterns across interviews which might be hardly considered as purely random, in particular if the repeated patterns show up concentrated on few interviewers. The application of their methodology points at "at risk" interviewers in a large German survey.

Menold, Winker, Storfinger and Kemper report on results from a research project aiming at combining several data based indicators to identify "at risk" interviews. Using experimental data, it is shown that the method has higher discriminatory power compared to approaches based on single characteristics. Furthermore, it might be more difficult if not impossible for an interviewer to match these traits for real data. I.e. even awareness of the method might not help to avoid detection.

The assessment of data based methods for identification of deviant behavior is hampered by the lack of datasets including identified falsifications. Typically, these are removed prior to making the dataset accessible. Thus, the methods presented in this part of the book can be evaluated only based on a quite limited number of test cases. In order to circumvent this restriction, Storfinger and Winker propose to use a simulation method, in the style of a bootstrap, to generate artificial datasets of predefined structure including both real and faked data. Then, the methods can be assessed on a large sample of test cases, and it becomes possible to identify features of the dataset which make the detection of deviant behavior more likely.

The second part of this volume is devoted to approaches from survey practice dealing with detection and prevention of deviant behavior.

Obviously, the use of interviewers in conducting surveys increases the risk of errors, changing behavior and uncontrollable interview situations. To minimize these risks standards and rules for interviewing are intended to guide the interviewers' behavior within the interview situation. These can be imparted by means of intensive training sessions and extensive training materials. In the first contribution of the second part, Jesske introduces the standards and standard methods used for interviewer training and monitoring at infas – the Institute for Applied Social Sciences. They are designed to flexibly meet the requirements of complex surveys for scientific purposes.

Market research studies have to be conducted within very short time frames, and should take into account the rising demands on quality standards. Under this postulates it is very important to have an easy-to-use tool to make interview control more efficient. In the second paper of this part Hülser describes a software called ALBERTA developed by GfK (Gesellschaft für Konsumforschung), which includes several tests that systematically check the quality of interviews and interviewers. By means of statistical criteria ALBERTA systematically identifies conspicuous interviews and interviewers.

Massing, Ackermann, Martin, Zabal and Rammstedt report on the specific and extensive quality standards for PIAAC, an international comparative survey initiated by the OECD (Organization for Economic Co-operation and Development), which aims at the assessment of adult skills and competencies to develop an empirically-based foundation for political interventions. The article focuses on the international quality standards related to interviewer control and their implementation in Germany. In addition the interviewer trainings conducted in Germany are described as well as the potential of such trainings in preventing interview falsification.

In the last contribution of the second part, Slavec and Vehovar present methods to detect irregularities in surveys caused by interviewer fabrication of questionnaires or other types of deviations from designed guidelines and instructions in the Slovenian National Readership Survey (NRS). As the authors point out, re-contact procedures are the most effective strategies to detect such irregularities. The results of the re-contact survey in the Slovenian NRS show that only a few respondents were not really interviewed, and the accordance of answers with the survey database is quite high, but for some interviewers, a higher rate of anomalies was perceived. In addition, analysis of non-respondents indicated that most of them are eligible and could be interviewed with more resources invested in fieldwork efforts to prevent interviewers from omitting cases.

The final part of the book on "Discourses on interviewers' behavior and deviations in survey data" includes papers which discuss different facets of deviations in surveys.

In the first contribution, Brüderl, Huyer-May and Schmiedeberg demonstrate interviewers' impact on the collection of data related to social networks in a German Family Panel. The authors find a strong variation across interviewers in producing rather large or rather small social networks, which is neither related to the characteristics of respondents nor to those of interviewers.

The paper by Menold and Opper refers to the work by Menold, Winker, Storfinger and Kemper as well as by Storfinger and Winker in the first part of this volume. The authors test the assumption that professional experienced interviewers would produce less pronounced differences from the real data than falsifiers without professional experience. The results have implications for detection methods described in the first part of the book. They are of particular interest given that previous studies on the "quality" of faked data were conducted using mainly students, i.e. inexperienced intervieweres, as falsifiers.

Haan, Ongena and Huiskes focus on the effects on data quality when interviewers deviate from the instruction while conducting interviews. Are these effects always negative? The authors classify interviewers' deviations in terms of their function in a conversation situation. As a result, the potential positive impact of interviewers' deviations is stressed. For example, interviewers might deviate to cope with difficulties during the interview in reaching cohesion and coherence as conversation principles. However, some interviewers' deviations are also associated with lower data quality. The authors discuss how such deviations could be avoided in surveys, e.g. by taking into account conversation rules when developing questionnaires.

The next chapter presents a conceptual work by Gwartney, in which interviewers' functions in surveys are embedded in the context of total survey error. The author differentiates between deliberate deviations from prescribed standards and not deliberate mistakes. Based on a discussion on interviewers' motivation to deviate (or to produce high quality survey data) Gwartney presents ways of reducing interviewers' deviations and mistakes. In doing so, she stresses the impact of organizational culture which acknowledges the crucial role of interviewers in surveys.

In the last contribution of part three by Ziegler and Kemper, not the faking behavior by interviewers, but that by the respondents is addressed. In particular, for the detection methods which are based on an analysis of existing survey data – presented in the previous parts of the book – it is important to differentiate between two sources of poor data quality: interviewers' deviations and respondents faking behavior. The authors discuss the mixed Rasch model as a chance to correct for the respondents' response bias which results from their faking behavior.

Outlook

The contributions in this book provide an international collection of recent research on interviewers' deviant behavior. It provides new evidence for the prevalence of interviewers' deviant behavior including some striking examples. However, it also shows how careful planning and supervision might help to avoid such unwelcome deviations from survey standards. The conclusions drawn by the authors are relevant both for field organizations organizing surveys and for researchers who have to assess the quality of collected data. Some methods can be applied to check for potential deviant behavior, other described approaches might proof helpful in avoiding such deviant behavior.

However, three issues become also evident. First, there remain a substantial number of unanswered questions regarding interviewers' motivation to deviate and their behavior when deviating. Several contributions highlight fields deserving future research. Thus, we hope to see more theoretical analysis and empirical evidence on these topics in the near future. Second, literature review and new cases discussed in this volume indicate that deviant behavior may be more widespread than the lack of reporting in most survey based research seems to suggest. Unfortunately, it appears much less obvious on how to collect reliable empirical evidence on the extent and impact of the phenomenon in actual scientific and applied research. Again, further research might be devoted to find ways for closing this gap. Third, given the substantial impact falsifications in survey data might have on further empirical research, scientific conclusions and policy recommendation, it appears imperative to find ways and develop methods for improving the quality of survey data by reducing the risk of such falsifications.

During the workshop, in a round table discussion, several ways to increase the incentives for all involved actors to care about survey quality and to report about deviant behavior have been discussed. One key issue identified was reputation. While reputation plays already an important role at the level of survey institutes, it is more difficult for interviewers to build up reputation. In fact, they might gain reputation while working for a particular survey institute, but will not be able to transfer this reputation when chancing affiliation as it is not documented anywhere expect in the organization's memory. Hence, it has been discussed whether assigning a unique interviewer identifier registered at a neutral research funding agency might help to overcome this problem. Then, registered interviewers might receive an extra bonus, while researchers could ask to involve only such registered interviewers. Furthermore, if some deviant behavior becomes apparent, the registering institution could inform all researchers having used the respective interviewer about it. However, such an infrastructure would come at some cost, and it might be questionable whether researchers have strong enough incentives to spend more on their surveys in order to reduce the risk of interviewers' deviant behavior. In fact, as some contributions in this volume demonstrate, one of the key drivers of deviant behavior is payment of interviewers based on completed interviews. Thus, changing only the payment structure might be already a first step towards setting the incentives right. Infrastructure for building up reputation and for eliminating doubtful data later when misconduct becomes apparent might follow.

Obviously, this introduction to our volume is not the place for a conclusive discussion of possible remedies, but we would like to stress again the importance of doing all possible to obtain high quality data not affected by interviewers' deviant behavior. We look forward to further research on the issue and alternative suggestions on how getting the incentives right.

Table of Contents

I Methods for Identification of Falsifications

A Literature Review of Methods to Detect Fabricated Survey Data

Sebastian Bredl, Nina Storfinger, Natalja Menold

Abstract

This paper reviews literature dealing with the issue of detecting interviewers who falsify survey data. The most reliable method of detecting falsifiers is through face-to-face reinterviewing of survey participants. However, especially in large scale surveys only a limited number of participants can usually be reinterviewed. A review of the present literature clearly indicates that reinterviewing is more effective if the reinterview sample is based on some indicators that might comprise metadata, survey data, or interviewer characteristics. We examine relevant literature with regard to the suitability of different types of indicators that have been used in this context.

Keywords: Interviewer falsification, quality control of survey data, reinterview

Acknowledgements

Financial support through the DFG in project WI 2024/2-1; ME 3538/2-1 within SPP 1292 is gratefully acknowledged. Furthermore we would like to thank Andreas Diekmann, Gesine Güllner and Peter Winker for their valuable comments on previous versions of the paper.

Introduction

In economic and social research, survey data is often the cornerstone of empirical investigations. Several factors that may impair the quality of such data during the period of field work, such as systematic non-response or interviewer effects on response behaviour, have gained attention in literature. Another important factor that has not received as much attention thus far is the conscious deviation from prescribed procedures by the interviewer, which is referred to as interviewer falsification (Schreiner et al., 1988) or cheating (Schräpler and Wagner, 2003). The American Association for Public Opinion Research (AAPOR) defines behaviour like this as 'intentional departure from the designed interviewer guidelines and instructions, unreported by the interviewer, which could result in the contamination of data' as 'interviewers' falsification' (2003: 1). There is a wide range of potential forms of cheating (cf. also Schräpler, 2010). The most blatant of these is undoubtedly the fabrication of entire interviews without ever having contacted the target person. Another possibility is partial fabrication, for example by making the contact but only asking some of the questions contained in the questionnaire and faking the remaining data (Harrisson, 1947). More subtle forms are listed by Case (1971), who mentions interviewing someone other than the intended person, changing the interview mode, or changing the location of the interview. The present chapter reviews literature dealing with detecting fabrication of complete interviews as the most blatant form of cheating as well as literature dealing with the detection of partial falsification.

Seen from the interviewer's perspective, there are several reasons why data fabrication might be an attractive option. Interviewers do not usually have a strong interest in delivering high-quality data, apart from the potentially satisfying feeling of having done a good job. As Durant (1946: 290) puts it, '[o]ne day's interviewing, however well done, merely serves to lead on to the next day's interviewing'. Furthermore, interviewers have to ask people who they do not know to reveal personal information, which may trigger dismissive reactions (cf. Crespi, 1945, Stewart and Flowerman, 1951, Köhne-Finster and Güllner, 2009) and are often faced with payment schemes based largely on the number of completed interviews (Kennickell, 2002). This might create pressure to augment 'quantity' and neglect the 'quality' of interviews, and may finally promote conditions leading to data fabrication (cf. Bennett, 1948, Sudman, 1966).

So far very little research has been done on the consequences of data fabrication for subsequent statistical analyses. This might be due in part to the fact that the severity of these consequences is obviously related to the prevalence of data fabrication. This parameter can be estimated only roughly, as it is likely that not all relevant cases can be detected. Studies reporting some estimates (e.g. Schreiner et al., 1988, Koch, 1995, Krejsa et al., 1999, Schräpler and Wagner, 2005, Li et al., 2009) suggest that the proportion of fabricated interviews rarely exceeds 5%. However, these studies refer only to large-scale surveys. In smaller surveys, with only a handful of interviewers, one may observe much larger proportions of fabricated interviews (Harrison and Krauss, 2002, Bredl et al, 2012). Not only is the quantity of fabricated data an important determinant in this context, but so is quality. If cheaters were able to reproduce "realistic" data, there would hardly be a problem. According to several studies (Hippler, 1979, Reuband, 1990, Schnell, 1991, Schräpler and Wagner, 2005), cheaters generally do quite a good job of fitting their data to marginal distributions found in real data, but they struggle to reproduce more complex relationships like those revealed by factor analysis or multivariate regression analyses. Consequently, even a small proportion of fabricated interviews, say of around five percent, might have a severe impact on the results of multivariate statistical analysis as shown by Schräpler and Wagner (2005). But this is not necessarily the case as demonstrated by Schnell (1991).

As interviewer data fabrication seems to be a non-negligible problem, one must be concerned about how to detect fraudulent interviews. Although the overall volume of literature on this issue is still modest, the variety of proposed methods and indicators is quite considerable, which clearly calls for some comparison and evaluation of different approaches. This is the issue we would like to address in this literature review. Based on our analyses we also try to formulate some recommendations on how to proceed in order to detect fabricated data, and we identify fields of research that need more attention in the future.

For our literature review we systematically searched different data bases for the social and economic sciences. Thereby, we analysed literature, published in English and German. Of the literature found dealing with complete or partial fabrication of interviews, the majority concerned methods of detecting falsifiers (most were journal articles, but conference proceedings and working papers were also available). In our review, we considered contributions on methods of detection based on empirical data. Overall, our search results show that up to now no extended research exists on the topic of falsifications. Nevertheless, we were able to find interesting results with respect to detection methods and to discuss the advantages and disadvantages of the different methods.

In the section "Overview of Key Studies", we examine five key studies which either applied detection methods during the field control in order to identify falsifiers (ex-ante studies) or tested the performance of several methods using datasets with known cases of falsification (ex-post studies). The aim of ex-post studies is to identify indicators that differ for data collected honestly and data which has been falsified. Based on this examination, the section "Overview of Different Approaches" discusses different methods for detecting data fabrication. Here, we focus on the effectiveness and the generalisability of the respective method. "Discussion and Outlook" summarizes the findings of our literature review and formulates some recommendations based on insights from the previous sections. Furthermore, this section highlights fields in which more research is needed.

Overview of Key Studies

In this section, we characterise selected comprehensive studies dealing with the detection of fabricated data. Table 1.1 provides an overview of these studies. As mentioned above, we distinguish between ex-ante studies employing the respective methods in order to detect falsifiers and ex-post studies that tested several indicators in datasets with known cases of data fabrication. All ex-ante studies included in the table used recontact procedures combined with other methods. With respect to the proportion of fabricated interviews we provide two numbers for ex-ante studies: the first refers to the proportion of falsified interviews in a random recontact sample, the second to the proportion obtained when recontact procedures were combined with other methods. Within ex-ante and ex-post studies different data analyses were conducted, using meta-data or collected survey data. Metadata, also called para-data, are survey process data, such as contact outcomes, obtained by interviewers or data produced during the interview (e.g. with the help of time stamps). Other analyses of survey data include a comparison of answers to survey questions, response sets (or response behaviour), and the application of Benford's Law.

Table 1.1: Selected studies dealing with the detection of data fabrication

Authors	**Survey**	**Share fabricated Interviews**	**Detection methods:**			
			Recontact	Metadata	Benford's Law	Other Analyses
		Ex-ante studies				
Koch (1995)	Large scale survey; German population; ALLBUS	random: 0.4%, combined: 2.3%	**X**			**X**
Hood, Bushery (1997)	Large scale survey; US poulation; NHIS	random: 0.2%, combined: 3.6%	**X**	**X**		**X**
Turner et al. (2002)	Large scale survey; Baltimore population	49% of 451 interviews contributed by 6 falsifiers (in total: 1200 interviews)	**X**	**X**		
		Ex-post studies				
Murphy et al. (2004)	Large scale survey; US population; NSDUH	19.5% in one highly affected US-state, no information on other states		**X**		**X**
Schraepler and Wagner (2005)	Large scale survey; German population; GSEOP	Sample A:0.6%; Sample B: 1.5% Sample C: 2%			**X**	**X**

Koch (1995)

Koch (1995) describes control procedures and their results in a survey of the German population (ALLBUS, German General Social Survey, 1994). In 1994 personal registers from registration offices started being used in the ALLBUS as sample frame. The previous sample method was 'random route' (ADM-System, Heyde and Loeffler (1993)), in which interviewers selected sample units within the two last stages of the selection process.

In contrast to ADM-samples, selected persons in personal register samples were known prior to data collection. Additionally, information about gender and age of sampled persons was provided in the sample frame. Interviewers received names and addresses of selected persons and should have interviewed exactly these persons. Hence, in the ALLBUS 1994 Koch (1995) was able to systematically check for falsifications by comparing the information on gender and age in the survey data with the data from the registration offices. Overall, the control procedures combined different steps:

A portion of interviews (25%) was routinely controlled by the survey institute responsible for data collection using postcards – they obtained a 60% response rate. These controls found 15 cases which were conducted incorrectly. Hence, these controls did not reveal considerable information about problems with the data.

In addition, all 3505 interviews realised in the ALLBUS 1994 were controlled by Koch, comparing gender and age of selected and interviewed persons. All cases with deviations detected by Koch (n = 196) were controlled by a new contact (in person, by phone or by post). Fraudulent interviewer behaviour could be detected in 81 cases (2.3%), of which 45 were complete falsifications of the interview. Koch emphasizes that the detection method he used in the ALLBUS is restricted by the sample method used. Samples, which use one or more selection stages, in which interviewers are involved (random route or samples with address registers as sample frame), cannot effectively apply this method, since the selected person is – as a rule – unknown prior to data collection. Another restriction of this method is that age and gender provide insufficient information to effectively expose falsified interviews. In most cases gender is easy to determine by the target person's first name, and age could be estimated by interviewers or asked in a short interview with the target person or with other household members (even with neighbours). The use of age and gender as information can allow only for the detection of significant carelessness in interviewers' work or other technical problems in the field, for example. It seems plausible to assume that falsifiers who are more cautious are not detected by the procedure described by Koch. Thus, the level of 2.3% of detected falsifications represents a lower bound for very crude fabrications. Nevertheless, Koch's work indicates that a more focused recontact procedure is more effective than controls conducted by the survey institute with a portion of interviewed persons who are selected without deliberate considerations.

Hood and Bushery (1997)

Hood and Bushery (1997) investigated the usefulness of several indicators in order to create a focused reinterview sample that could be applied to the US-National Health Interview Survey (NHIS). According to the authors data fabrication occurs rarely in the NHIS. As a result, many reinterviews are required to detect a falsifier. In this context the authors emphasize the usefulness of a focused reinterview that concentrates on interviewers who seem to be more likely than others to have fabricated data according to some indicators.

Hood and Bushery assume that cheating interviewers try to 'keep it simple' (p. 820). Thus, they can be expected to label eligible households as ineligible and choose answers that allow questions to be skipped, leading to avoidance of subsequent optional parts of the questionnaire. For example, a considerable number of questions was not asked in white households in the NHIS. Consequently, a high proportion of white or ineligible households within an interviewer's assignment may be a sign of data fabrication.

The basic idea behind the approach is to examine data in questionnaires as well as some metadata (ineligible households) in order to identify interviewers who merit a closer look during the reinterview stage. However, it is clear that a relatively high proportion of white or ineligible households in one interviewer's assignments is not necessarily linked to dishonest behaviour, but rather might also be due to the specific characteristics of the area were the interviews were conducted. This is known as so-called spatial homogeneity (cluster related design effect; cf. Groves et al. (2004)), meaning in this case the homogeneity of individuals living within a geographical area. To differentiate between interviewer effects and spatial homogeneity, Hood and Bushery considered the differences between actual proportions and those that could be expected based on data from the 1990 census. If differences for all variables exceeded a certain threshold, the interviewer was flagged as an outlier and was then checked using focused reinterviews.

During the focused reinterview 3 falsifiers were detected from the 83 interviewers that were checked (3.6%). This 'success rate' is clearly above the 0.2% achieved by random reinterview. Although the informative value of these numbers should not be overrated, as they rely on a small number of cases, they do indicate that focused reinterviews deliver better results than purely random reinterviews.

The general problem with this approach is that discriminating between effects caused by data fabrication and those caused by the particularities of an interviewer's assignment is difficult. A reliable reference survey – like the 1990 census in the case of the Hood and Bushery study – is often simply not available. Furthermore – and a point also made by Hood and Bushery (1997) – in contrast to the study by Koch (1995) the approach considers interviewers and not interviewed individuals. This may be problematic if an interviewer fabricates only a small part of his assignments. In this case, indicators based on all interviews done by an interviewer might have only little discriminatory power.

Turner et al. (2002)

Turner et al. (2002) describe their painful experiences with falsifications of a large part of the sample in a Baltimore population survey. In contrast to national large scale surveys described above, this particular survey had two special aspects: firstly, it was related to a quite sensitive topic (sexually transmitted diseases) in which biological specimens were collected; secondly, it was a large local survey. This survey differs from national surveys for the second reason, since the latter does not need a large interview staff in a local area. It was particularly difficult for the data collection institute to recruit a sufficient number of interviewers in Baltimore. Turner et al. (2002) report that very low participation rates were obtained, and as a result additional interviewer trainings were conducted and the data collection period was extended.

The research team found irregularities in the data delivered by the data collection institute: six interviewers showed implausible success rates in conducting interviews. In fact 54% to 85% of assigned households were successfully interviewed by these interviewers, in contrast to other interviewers, who succeeded only in 31% of the cases on average. All interviews submitted by these six interviewers were verified by telephone or face-to-face recontact. In addition, controls for other interviewers were conducted. Here, the authors used metadata (cf. Table 1.1) to find suspect cases and combined them with a reinterview for verification. As a result it was found that 49% of the 451 interviews submitted by six suspected interviewers were falsifications.

The procedure by Turner et al. (2002) is similar to that reported by Koch (1995): research staff conducted controls independent of any controls conducted by the data collection institute. In contrast to Koch (1995), who checked only suspect cases, all interviews conducted by suspicious interviewers were controlled by Turner et al. (2002), with a high hit ratio for fabricated interviews. But in comparison to other studies, using the number of conducted interviews as a kind of metadata is restricted by the specifics of the survey. These specifics are associated with difficulties in conducting a local population survey on a sensitive topic. However, studies we discuss in this section show that local population surveys on sensitive topics are particularly prone to falsifications, and that it would be more effective to recontact all cases assigned to a dishonest interviewer.

Murphy et al. (2004)

Murphy et al. (2004) analysed data produced by three identified falsifiers in the American National Drug Survey on Drug Use and Health (NSDUH). This large scale survey selects around 70,000 persons each year who are interviewed using computer-assisted interviewing (CAPI) and audio computer-assisted self interviewing (ACASI), in which the laptop is given over to the respondent. Hence, the laptop registered time stamps for each question and each interview step in both modes, which allowed for the calculation of elapsed time for each respective action.

Like Turner et al. (2002) Murphy et al. (2004) examined response patterns to sensitive questions related to the lifetime use of cigarettes, alcohol, marijuana, cocaine and heroin. The authors calculated the proportion of respondents per interviewer who claimed to have already consumed the respective drug during their lifetime. To account for spatial homogeneity the authors controlled for demographic characteristics of the (alleged) respondents by examining shares separately for men and women, younger and older respondents and Hispanics and non-Hispanics. The resulting indicator performed extremely well in separating falsifiers and honest interviewers. In both cases, all three falsifiers were among the top four interviewers, if interviewers were ranked according to the values of an index indicating deviations between drug abuse rates in the interviewer's sample and the remaining data. As in the study by Turner et al. (2002) it turned out that falsifiers struggle to adequately reproduce answers to very sensitive questions.

Murphy et al. (2004) employed metadata – namely time stamps – in order to determine whether response times are different when falsifiers fabricate data as compared to situations in which the data is collected honestly. The NSDUH is a very interesting application in this regard, as it consists of the CAPI and the ACASI part. However, it turned out that clear patterns of differences between falsifiers and honest interviewers did not emerge for either the CAPI part or for the ACASI part. One of the falsifiers was generally much faster than the other interviewers, but the other two falsifiers were much slower.

Overall, the study suggests that responses to sensitive questions, rather than lengths of different interview modules, are a more reliable indicator for detecting cheating interviewers. When analysing these responses, it must be kept in mind that emerging patterns may also be due to the allocation of interviewers to certain areas. Considering this, it is important to somehow control for this factor. In contrast to the framework in which Hood and Bushery (1997) operated there was no reference study available to the authors, so they stratified their indicators according to some of the respondent's characteristics. Such an approach always bears the risk of omitting some decisive characteristics.

Schräpler and Wagner (2005)

Schräpler and Wagner (2005) examined the data from the German long-term panel study SOEP collected every year since 1984. In such panel studies fabrications are relatively rare since respondents are interviewed repeatedly and consistency checks across the different waves might immediately reveal fraudulent data. By means of two different ex-post analyses Schräpler and Wagner (2005) examined only data from the first waves of different subsamples of the SOEP (Schupp and Wagner, 2002).

The first one was based on the so called Benford's Law (Benford, 1938), which we illustrate in more detail in the same named section. The idea behind this method is to examine the distribution of the first digit of all numbers in the (metric) answers from the survey. If the numbers follow a specific monotonic declining distribution, simply spoken that the proportion of 1's is higher than the proportion of 9's, one can assume that the data is Benford distributed. Schräpler and Wagner (2005) calculated the deviation from the Benford distribution by means of a chi-square value for every interviewer cluster rather than for every single interview. As a result they showed that about half of the known fakers could indeed be marked as cheaters through application of this detection method.

Apart from this analysis Schräpler and Wagner (2005) created an interesting approach by considering several variables in order to detect known fakers in survey data. The authors called this the 'variability method' because the idea behind it is that cheaters show a lower variance of specific answers across all their conducted interviews than accurate interviewers do. Schräpler and Wagner (2005) attributed the reduction of variance to the proportions of missing answers, of extreme answers in scale questions and of conspicuously consistent answers across specific questions in the questionnaire. Based on the observed variance of interviews Schräpler and Wagner (2005) calculated a plausibility value for every interviewer in order to identify cheaters. If the plausibility was too low an interviewer was considered to be a faker. In this way the authors ranked interviewers with respect to their plausibility values and noticed that almost all of the known cheaters appeared at the top of the ranking. Additionally, they noted that their results were much better than those which are based on Benford's Law. Thus, it might be concluded from the results that the 'variability method' is a more promising way to reveal falsifiers than Benford's Law used as a stand alone indicator.

Résumé

Table 1.1 shows that random (or unfocused) recontacts enable detection of only an insignificant number of falsified interviews. In all ex-ante studies unfocused recontact procedures were then refined using supplemental information. This information was helpful in identifying additional fabricated interviews. However, the availability of supplemental information was highly dependent on the specific circumstances of the study. More concretely, sampling procedures, the survey topic, and the sensitivity of questions are all associated with the usability of a particular detection method. In principle, ex-post data analyses seem to be an effective method of identifying indicators in order to separate false and real data. But more research should be done in this area to assess the success of such methods, and it has to be stressed again that appropriate methods are often dependent on the specifics of a survey (e.g. the sensitivity of questions asked or the proportion of appropriate questions for the application of Benford's Law or the 'variability method').

Overview of Different Approaches

As outlined in the previous section existing literature suggests that the effectiveness of recontact procedures can be increased if they are combined with other indicators. In this chapter, we first examine literature related to recontact procedures and subsequently discuss the suitability of other methods that could be applied to create focused reinterview samples.

Recontact Procedures

The most common method of detecting faked survey data is the recontact method. Using this method respondents are recontacted in person, by mail or by telephone after the initial interview in order to verify whether the initial interview actually took place. Below we will focus on this recontact method and the possible problems associated with it.

In spite of the fact that AAPOR (2003) suggests that face-to-face recontact is the most effective method of detecting fraudulent data, for financial reasons the most common recontact method used in surveys involves sending postcards to interviewed persons with an appeal to them to reply. These postcards mainly ask respondents about the time, date, and critical components or topics of the interview, as well as the interviewer's behaviour. A statement about not being interviewed or implausible time and date information may then be considered as indicators of falsifications. In general this method has some questionable factors as shown by Koch (1995) and Hauck (1969). These factors include memory problems and the willingness of contacted persons to reply, and these are associated with a selectivity bias. In particular, with respect to response rates, telephone and/or personal contacts, also referred to as reinterviews, are more effective than contacts using postcards (Hauck, 1969).

Telephone or personal reinterviews were already mentioned in an early work by Case (1971). In market surveys he conducted controls with telephone reinterviews and revealed that about 27% of the interviews in all the studies examined were not conducted properly. Also, the U.S. Bureau of Census regularly checks a randomly selected portion of interviewers (between 2% and 10%) by reinterviewing particular target persons (cf. Bushery et al., 1999). But regarding response and memory problems the reinterview process is limited by sample size and duration (Cantwell et al., 1992). A large number of reinterviews increases costs and a high number of questions posed within a reinterview is a strain for respondents, and consequently biases results. Reinterviews are also limited by the elapsed time following the interview (a delayed survey for further control purposes (reinterview) bears the risk of memory effects). Thus, what is required is a reinterview sample which is large enough to generate significant results but is small enough to keep costs down.

As we have illustrated above (cf. section "Overview of Key Studies"), ex-ante studies show that a small, and most of all a non-randomly selected, reinterview sample is more effective in detecting cheating interviewers. In the following sections we would like to present the prevalent indicators used for the optimal creation of a focused reinterview sample: usage of metadata, Benford's Law and other statistical analyses of survey data, as well as interviewer characteristics.

Metadata

As already outlined in "Overview of Key Studies", the notion 'metadata' comprises different types of information related to the process of data collection, rather than to the collected data itself. Metadata-based indicators used to detect falsifiers can be divided into two groups: indicators based on interviewer's contact outcomes and indicators based on interview processing, such as date and time stamps.

Contact outcomes refer to information related to how many participants refused the interview or how many participants were ineligible for some reason. As outlined above, Turner et al. (2002) were able to detect a large number of fabricated interviews by focusing recontact efforts on interviewers who had shown a suspiciously high success rate, whereas Hood and Bushery (1997) employed the ineligible unit rate as an indicator to put together their focused reinterview sample. In this context it is recommended to control for the characteristics of the area where an interviewer conducts his/her work, as demonstrated by Hood and Bushery (1997).

Date and time stamps can only be recorded if the interview is conducted using a mode which relies on computer assistance. If these are available, they can be used to examine interview length, or the number of interviews completed within one day or within periods in which interviews were conducted. These types of indicators are employed by Bushery et al. (1999), Krejsa et al. (1999) and Murphy et al. (2004).

Given a very limited number of studies it has so far been quite difficult to evaluate how well metadata can be used to detect cheating interviewers. Krejsa et al. (1999) combined both types of metadata-based indicators, which delivered quite promising results. The results of Turner et al. (2002) are quite promising as well, although the results of Murphy et al. (2004) are less encouraging. In the case of Turner et al. (2002) it should also be kept in mind that general success rates were quite low, probably as a result of the high sensitivity of the questions asked in the course of interviews. Consequently, high rates for falsifiers were extremely noticeable.

A large advantage of approaches relying on metadata analysis is that they can be applied to a vast range of surveys. Whenever interviewers are prescribed which persons or households to contact, one can calculate indicators related to contact outcomes. Whenever interviews are conducted with computer assistance, there is the opportunity to record date and time stamps.

Benford's Law

The largest part of the scarce research regarding the usage of answer patterns to detect fraudulent interviews is related to the usage of Benford's Law for the analysis of metric survey data. With accurate survey data the distribution of the first digit of these metric answers usually follows the so called Benford's Law (Benford, 1938), a logarithmic (Newcomb, 1881) and scale invariant distribution (Hill, 1995). Thus, the probability that the first digit of the numbers is 1 is higher than the probability that it is 9. In general, Benford's Law could be adapted to data without a built-in maximum (Nigrini, 1999) and to data which is not composed of assigned numbers like zip codes or bank accounts. For example, Nigrini (1996) and Tödter (2007) have shown that business and financial data in particular follow this monotonic decreasing distribution.

In order to reveal faked survey data by means of Benford's Law one must ensure that accurate survey data is actually Benford distributed and that faked data is not. If we look at the literature concerning this topic we cannot assume that this is always true, as rounding effects might cause a high proportions of 0's and 5's as shown by Schräpler and Wagner (2003) and Wang and Pedlow (2005) who analyzed the first and second digits.

Schräpler and Wagner (2003) as well as Wang and Pedlow (2005), tested some modifications of the analysis in order to improve the detection of cheaters. Schräpler and Wagner (2003) calculated a chi-square value for 'interviewer clusters', meaning that they pooled together all interviews by one interviewer into one cluster. However, the clusters with faked data were not precisely revealed. Wang and Pedlow (2005) dealt with the rounding effect by using the 'all cases distribution' (cf. Swanson et al. (2003)). This approach consists in comparing the first digit distribution in one interviewers's questionnaires to the pooled distribution of all other interviewers instead of comparing it to the Benford distribution. By using this method instead of Benford's Law, it became possible for the authors to identify cheating interviewers.

Thus, we infer from the results of Wang and Pedlow (2005) and Schräpler and Wagner (2003) that real survey data does not always follow Benford's Law and that one must be aware of the occurrence of a rounding effect. Furthermore, Diekmann (2010) pointed out that one should avoid stating that survey data which deviate from Benford's Law are automatically falsified. It often occurs that the whole data set follows Benford's Law while some subsamples do not, even if they have not been falsified.

To summarize, the current work concerning Benford's Law questions the validity of this method. There is little empirical evidence that real survey data are close to the Benford distribution and problems like rounding effects may lead to a deviation from the Benford distribution. Thus, the 'all cases distribution' might be more appropriate. In addition, further modifications to Benford's Law, such as several combinations of digits, should be developed and tested in datasets with a higher proportion of falsifications.

Additionally, there is a lack of research regarding the number of interviewers and the number of interviews per interviewer which are required for the identification of falsifications to produce precise results. Recent work by Storfinger and Winker (2011), in which Benford's Law is used as one of four indicators, suggests that the performance somewhat deteriorates as the number of interviews per interviewer decreases and improves as the overall number of interviewers decreases. However, more research is needed to assess how many questionnaires per interviewer are needed in order to successfully adopt Benford's Law. Also, the number and types of variables in the questionnaire which are suitable for application to Benford's Law have not been precisely identified (cf. Porras and English, 2004). Is Benford's Law only appropriate for metric variables in raw survey data, or only for statistical estimates like regression coefficients? As Scott and Fasli (2001) demonstrated in a synthetic way, data are more likely to conform to Benford's Law if the dataset contains only positive numbers and is positively skewed with a modal value not equal to zero. But there is still a lack of evidence for real survey data which fit to the Benford distribution. So far Benford's Law can be used to evaluate survey data quality (cf. Judge and Schechter, 2007) but it appears to be not efficient enough to precisely identify cheaters.

Other Statistical Analyses of Survey Data

Inspection of concrete survey data also delivers encouraging results by revealing 'at risk' interviewers and falsifications. Several forms of questions and questionnaires could be included in such an analysis.

As a result of our literature review we differentiate between two kinds of indicators which can be applied to an analysis to compare false and real data. The first kind applies to more or less plausible answers to survey questions by falsifiers, while the second kind is associated with the answering behaviour of falsifiers, which may differ from that of real respondents. We will call the first kind of indicators 'content related' and the second 'formal'.

Content related indicators are content related answer patterns which systematically differ between fraudulent and accurate collected data. In the study by Turner et al. (2002) target persons 'interviewed' by falsifiers 'reported' having remarkably more lifetime partners and 'showed' substantially more sexual activity. Finally, falsifiers provided phone numbers of 'interviewed' persons less often. Further examples – as found by Reuband (1990) – are differences in evaluating one's own personal financial situation, which was estimated more optimistically by falsifiers than by real sampled persons. However, content related indexes have seldom been used in studies to detect falsified data. Schnell (1991), who systematically analysed differences between real and falsified data and in particular found differences in correlations and multiple regressions, remains sceptical about the usage of obtained differences for the purpose of detection. Of course, applying content criteria, apart from questions about household composition, age and gender, is problematic since it requires asking sensitive questions (e.g. Turner et al., 2002) or questions about content not related to the survey topic. Application of content related indicators needs additional strong hypotheses and knowledge about the differences between falsified and real data, which are, as a rule, not available.

Formal criteria are produced by analysing the answering behaviour of interviewed persons. An example is item non-response. As shown by Schräpler and Wagner (2005) and Bredl et al. (2012) this occurs less often in falsified than in real interviews, since falsifiers incorrectly assume that real respondents would answer all questions in an interview – and, therefore, they avoid item nonresponse. In addition, Bredl et al. (2012) showed that falsifiers differ from real survey respondents as they avoid to denominate the category 'others' in semi-open-ended questions and that they are less extreme if they use rating scales for their answers. With the help of multivariate cluster analyses, Bredl et al. (2012) were able to separate falsifiers from honest interviewers using (in addition to Benfords Law, see same named section) information on item non-response, extreme answers and answering of open and semi-open questions. Furthermore, cheating interviewers can also be identified by looking at the proportion of answers to questions that lead to a faster answering, and therefore a quicker interview (cf. Matschinger et al., 2005, Hood and Bushery, 1997). For a health survey in the US, Hood and Bushery (1997) reported that falsifiers selected 'the shortest path through the interview' producing survey participants who live in one person families and are 'white non-smoker, no health problems and no health insurance' (p. 821).

In summary, formal criteria have advantages over content criteria in that they can be used for different types of content in different surveys. Work by Schräpler and Wagner (2005) and Bredl et al. (2012) shows that using formal indicators is an encouraging approach. Like in the case of content related criteria, the detectors should have some ideas on falsifiers' answering behaviour. Furthermore, it is important to ensure that falsifiers cannot adapt their way of cheating easily if the criteria are known to them (e.g. if a falsifier is aware that interviewers are considered to be suspicious if they deliver questionnaires with a low prevalence of item non-responses he/she could simply increase the prevalence of item non-responses when fabricating the data). This problem might be tackled by combining several criteria which makes adaptation from the falsifier's side more difficult. Altogether, more research is needed in order to find an optimal questionnaire form, which can help to detect falsifiers through statistical data analysis using formal criteria.

Interviewer Characteristics

In most of the studies presented above interviewer characteristics were discussed as one issue which can be used for more focused reinterviews. The authors agree that inexperienced interviewers are more likely to show cheating behaviour and should consequently be controlled by extended procedures (Biemer and Stokes, 1989, Wetzel, 2003, Turner et al., 2002, Schreiner et al., 1988). Schreiner et al. (1988) recommended that '(...) for the newer interviewers it may be useful to reinterview some of their work more frequently' (Schreiner et al., 1988: 496). Concerning other interviewer characteristics mixed results can be found in the literature. With respect to gender and age of interviewers Koch (1995) has shown that young interviewers with a higher level of education produce a higher rate of falsifications. In West-Germany, Koch found no gender differences, but in East-Germany male interviewers fabricated interviews more often. However Schräpler and Wagner (2003) did not discover any age, gender or education effects in SOEP.

As a result, interviewers' length of service seems to be a sufficient cue for more focused reinterviews. However, experienced interviewers might not be less likely to falsify, but just less likely to be detected by controls. Hood and Bushery (1997) reported the results of a study by the US Census Bureau 1982 which analysed interviewers' characteristics and the likelihood of falsification. The results showed (similar to other studies) that interviewers' tenure in particular is associated with differences in falsification behaviour. New interviewers with less than five years of experience had a higher probability of being detected, since they falsify more of their assignments and they tend to falsify entire interviews. Experienced interviewers falsify a smaller proportion of their assignments and prefer to falsify only a part of the interview. As a consequence, falsifications by experienced interviewers are more difficult to detect. Additionally, as shown by Schreiner et al. (1988) experienced interviewers use more selective approaches for falsification than less experienced interviewers. In panel surveys, for example, experienced interviewers falsify more often in continuing households (and not in newly selected households), since data can be appropriately estimated from the previous round.

However, in particular the comparison between experienced and inexperienced interviewers is based on very small sample sizes of interviewers who were found to falsify, and thus analyses were often done without statistical tests or results were afflicted by high statistical uncertainty.

In summary, newly hired and relatively inexperienced interviewers are seen as requiring extensive controls, but at the same time they are more amateur falsifiers who are more likely to be detected. This topic needs more research, especially regarding methods which can help to identify partially fabricated interviews conducted by experienced interviewers. Using length of service or other demographic characteristics for generating focused reinterviews appears to be less efficient in light of the results discussed in this section. This is particularly due to the fact that extended controls of less experienced interviewers lead to lower rates of falsification detection for experienced interviewers, who are much more sophisticated falsifiers.

Discussion and Outlook

In spite of the scarcity of scientific publications related to detection of data fabrication by interviewers, an examination of existing literature has delivered useful insights. Below we attempt to deduce some recommendations for practitioners.

We have distinguished between two types of studies: ex-ante studies describe approaches implemented to detect cheaters, whereas ex-post studies apply indicators to datasets with known cases of falsifications. Ex-ante studies analysed in the section "Overview of Key Studies" clearly suggest that focused recontacts are more effective than recontacts based on random samples. Furthermore, examination of studies using different recontact procedures reveals that the reinterview is the most reliable of these procedures. We thus consider the focused reinterview as a good strategy to detect falsifiers.

Ex-post studies complement ex-ante studies in that they provide deeper insights into the suitability of different methods when creating focused reinterview samples. We have defined four methods in this context: metadata, Benford's Law, other analyses of data contained in questionnaires referring to formal or content-related information, and interviewer characteristics. These can be divided into methods available for most surveys and methods whose applicability is limited to specific types of surveys.

Metadata, most types of formal survey data and data on interviewer characteristics are available for a wide range of surveys. Approaches relying on metadata and formal survey data have delivered promising results when used to detect falsifiers. This is especially true in the case of contact outcome data. Analysis of answer patterns for filter questions seems to be another promising approach and is directly linked to hypotheses about falsifiers' behaviour. Interviewer characteristics did not turn out to be a useful indicator for creating focused reinterview samples. However, literature reveals one important point: experienced interviewers falsify in ways that make their detection more difficult. Thus, indicators relying on metadata or survey data can be expected to deliver better results for inexperienced interviewers who, if they falsify data, do it in a more amateur way. Consequently, it is plausible to assume that focused reinterviews are less efficient when applied to experienced interviewers.

The applicability of Benford's Law and of content related indicators based on specific answers by survey participants depends on the nature of the data collected. Using Benford's Law requires a wide range of metric variables. Even if a multitude of these variables is available it still remains open whether data which is honestly collected can be supposed to conform to Benford's Law, and whether fabricated data cannot. There have been some promising results, but Benford's Law should be applied with caution and, if possible, in combination with other indicators. General statements about the usefulness of content-related survey data are difficult to make, as the type of data available depends on the type of survey. It can be stated that answer patterns to questions on very sensitive issues have turned out to be good predictors of interviewer cheating and should thus be used to detect cheaters when available.

Whenever one employs indicators based on metadata or content-related survey data one has to keep in mind that striking indicator values are not necessarily caused by data fabrication but may also be the result of 'conventional' interviewer effects or cluster related design effects (spatial homogeneity). Several studies which attempt to control for spatial homogeneity have been discussed, however these approaches are not always replicable and they bear the risk of omitting decisive factors. It is possible that indicators based on formal survey data – for example the proportion of extreme answers or item non-response – are less impaired by conventional interviewer effects and spatial homogeneity. However, to the best of our knowledge no research has yet been done on this issue.

In addition to the recommendations we have provided above, we would like to point out some ideas for further research. First of all, more research should be done to broaden the discussed findings with respect to the suitability of methods we presented in "Overview of Different Approaches". Of course there are several studies which focus on one or more of these methods, but as we have shown in the context of Benford's Law the results are sometimes inconclusive or are based on a sample which is too small. Thus, further studies are needed to gain more reliable insights into how every single detection method performs, as well as how combinations of these methods perform.

In particular, the usage of survey data, and especially several content-related and formal indicators, poses the problem of a trade-off between the complexity of questionnaires and their usability on the part of respondents. Following our recommendation that one should apply different detection methods, the questionnaire to be examined should provide a high proportion of suitable questions to derive several indicators for the statistical analyses of survey data. In doing so, the complexity of the questionnaire may increase and therefore the usability for the interviewer, and of course for the respondent, diminishes and produces further biases. Thus, one must keep in mind that the questionnaire overburdens neither interviewers nor respondents.

Concerning the structure of survey data examined and regardless of the detection method used, we noticed that in the majority of the studies presented, analyses were conducted only on the interviewer level. Yet identification of fraudulent survey data could also be realized on the interview level (cases in a data set). Focusing on interviews bears the problem that the amount of data per interview is less than the amount of data per interviewer, which probably reduces discriminatory power. On the other hand, approaches focusing on interviewers might struggle to detect falsifiers who fabricate only a small proportion of their interviews. Which strategy is preferable is another topic for future research. Finally, it should be kept in mind that the methods discussed above have all been tested and developed primarily to identify fabrications of entire interviews. Further work concentrating mainly on the identification of partially faked interviews is also needed.

The literature overview presented here shows that first conclusions can be drawn from existing comprehensive studies about interview falsifications and the prevailing detection methods. However, most of the approaches – especially in the analysis of collected data – should be evaluated through further research related to their effectiveness, and approaches should be developed to enable an improved detection of falsifications and falsifiers.

Bibliography

AAPOR (American Association for Public Opinion Research): Interviewer Falsification in Survey Research: Current Best Methods for Prevention, 2003, Detection and Repair of its Effects. Available under: http://www.aapor.org/Content/NavigationMenu/ResourcesforResearchers/falsification.pdf (Access: 18.11.2010).

Benford, Frank: The Law of Anomalous Numbers, Proceedings of the American Philosophical Society, 1938, 78 (1), 551–572.

Bennett, Archibald S.: Toward a Solution of the 'Cheater Problem' Among Part-Time Research Investigators, Journal of Marketing, 1948, 12 (4), 470–474.

Biemer, Paul and Stokes, S. Lynne: The Optimal Design Quality Control Sample to Detect Interviewer Cheating, Journal of Official Statistics, 1989, 5 (1), 23–29.

Bredl, Sebastian, Winker, Peter and Kötschau, Kerstin: A Statistical Approach to Detect Interviewer Falsification of Survey Data , Survey Methodology, 2012, 38 (1), 1-10.

Bushery, John M., Reichert, Jennifer W., Albright, Keith A. and Rossiter, John C.: Using Date and Time Stamps to Detect Interviewer Falsification, in: Proceedings of the American Statistical Association (Survey Research Methods Section), 1999, 316–320.

Cantwell, Patrick J., Bushery, John M. and Biemer, Paul P.: Toward a Quality Improvement System for Field Interviewing: Putting Content Reinterview Into Perspective, in: Proceedings of the American Statistical Association (Survey Research Methods Section), 1992, 74–83.

Case, Peter B.: How to Catch Interviewer Errors, Journal of Advertising Research, 1971, 11 (2), 39–43.

Crespi, Leo P.: The Cheater Problem in Polling, Public Opinion Quarterly, 1945, 9 (4), 431–445.

Diekmann, Andreas: Benford's Law and Fraud Detection. Facts and Legends, ETH Zurich Sociology Working Paper No. 8, 2010.

Durant, Henry: The 'Cheater' Problem, Public Opinion Quarterly, 1946, 10 (2), 288–291.

Groves, Robert M., Fowler Jr., Floyd J., Couper, Mick P., Lepkowski, James M., Singer, Eleanor and Tourangeau, Roger, Survey Methodology, New Jersey, 2004, Wiley.

Harrison, David E. and Krauss, Stefanie I.: Interviewer Cheating: Implications for Research on Entrepreneurship in Africa, Journal of Developmental Entrepreneurship, 2002, 7 (3), 319–330.

Harrisson, Tom: A British View on 'Cheating', Public Opinion Quarterly, 1947, 11 (1), 172–173.

Hauck, Mathew: Is Survey Postcard Verification Effective?, Public Opinion Quarterly, 1969, 33 (1), 117–120.

Heyde, Christian von der and Loeffler, Ute: Die ADM-Stichprobe, Planung und Analyse 5: 49–53.

Hill, Theodore P.: A Statistical Derivation of the Significant Digit Law, Statistical Science, 1995, 10 (4), 354–363.

Hippler, Hans J.: Untersuchung zur Qualität absichtlich gefälschter Interviews, ZUMA Arbeitspapier, 1979.

Hood, Catherine C. and Bushery, John M.: Getting more Bang from the Reinterviewer Buck: Identifying 'At Risk' Interviewers, in: Proceedings of the American Statistical Association (Survey Research Methods Section), 1997, 820–824.

Judge, George and Schechter, Laura: Detecting Problems in Survey Data Using Benford's Law, Journal of Human Resources, 2007 44 (1), 1–24.

Kennickell, Arthur B.: Interviewers and Data Quality: Evidence From the 2001 Survey of Consumer Finances', in: Proceedings of the American Statistical Association (Survey Research Methods Section), 2002, 1807–1812.

Koch, Achim: Gefälschte Interviews: Ergebnisse der Interviewerkontrolle beim ALLBUS 1994, ZUMA-Nachrichten, 1995, 36, 89–105.

Köhne-Finster, Sabine and Güllner, Gesine: Ergebnisse der Interviewerbefragung im Mikrozensus, Wirtschaft und Statistik, 2009, 5, 397–405.

Krejsa, Elizabeth A., Davis, Mary C. and Hill, Joan M.: Evaluation of the Quality Assurance Falsification Interview Used in the Census 2000 Dress Rehearsal, in: Proceedings of the American Statistical Association (Survey Research Methods Section), 1999, 635–640.

Krosnick, Jon A. and Alwin, Duane F.: An Evaluation of a Cognitive Theory of Response-Order Effects in Survey Measurement, Public Opinion Quarterly, 1987, 51 (2), 201–219.

Li, Jianzhu, Brick, J. Michael, Tran, Bac and Singer, Phyllis: Using Statistical Models for Sample Design of a Reinterview Program, in: Proceedings of the American Statistical Association (Survey Research Methods Section), 2009, 4681–4695.

Matschinger, Herbert, Bernert, Sebastian and Angermeyer, Matthias C.: An Analysis of Interviewer Effects on Screening Questions in a Computer Assisted Personal Mental Health Survey, Journal of Official Statistics, 2005, 21 (4), 657–674.

Murphy, Joe, Baxter, Rodney, Eyerman, Joe, Cunningham, David and Kennet, Joel: A System for Detecting Interviewer Falsification, Paper Presented at the American Association for Public Opinion Research 59th Annual Conference, 2004.

Newcomb, Simon: Note on the Frequency of Use of the Different Digits in Natural Numbers, American Journal of Mathematics, 1881, 4 (1/4), 39–40.

Nigrini, Mark J.: A Taxpayers Compliance Application of Benford's Law, Journal of the American Taxation Association, 1996, 18, 72–91.

Nigrini, Mark J.: I've got your Number, Journal of Accountancy, 1999, 187 (5), 79–83.

Porras, Javier and English, Ned: Data-Driven Approaches to Identifying Interviewer Data Falsification: The Case of Health Surveys, in: Proceedings of the American Statistical Association (Survey Research Methods Section), 2004, 4223–4228.

Reuband, Karl-Heinz: Interviews, die keine sind. ‚Erfolge' und ‚Mißerfolge' beim Fälschen von Interviews, Kölner Zeitschrift für Soziologie und Sozialpsychologie, 1990, 42 (4), 706–733.

Schnell, Rainer: Der Einfluss gefälschter Interviews auf Survey Ergebnisse, Zeitschrift für Soziologie, 1991, 20 (1), 25–35.

Schräpler, Jörg-Peter: Benford's Law as an Instrument for Fraud Detection in Surveys Using the Data of the Socio-Economic Panel (SOEP), SOEP Papers, 2010.

Schräpler, Jörg-Peter and Wagner, Gert G.: Identification, Characteristics and Impact of Faked Interviews in Surveys – An analysis by means of genuine fakes in the raw data of SOEP, IZA Discussion Paper Series, 969, 2003.

Schräpler, Jörg-Peter and Wagner, Gert G.: Characteristics and Impact of Faked Interviews in Surveys – An Analysis of Genuine Fakes in the Raw Data of SOEP, Allgemeines Statistisches Archiv, 2005, 89, 7–20.

Schreiner, Irwin, Pennie, Karen and Newbrough, Jennifer: Interviewer Falsification in Census Bureau Surveys, in: Proceedings of the American Statistical Association (Survey Research Methods Section), 1988, 491–496.

Schupp, Jürgen and Wagner, Gert G.: Maintenance of and Innovation in Long-Term Panel Studies: The Case of the German Socio-Economic Panel (GSOEP), Allgemeines Statistisches Archiv, 2002, 86 (2), 163–175.

Scott, Paul and Fasli, Maria: Benford's Law: An Empirical Investigation and a Novel Explanation, CSM technical report, Department of Computer Science, University Essex, 2001.

Stewart, Naomi and Flowerman, Samuel H.: An Investigation of Two Different Methods for Evaluation of Interviewer Job Performance, Personnel Psychology, 1951, 4 (2): 161–170.

Storfinger, Nina and Winker, Peter: Robustness of Clustering Methods for Identification of Potential Falsifications in Survey Data, ZEU Discussion Paper Nr. 57, 2011.

Sudman, Seymour: New Approaches to Control of Interviewing Costs, Journal of Marketing Research, 1966, 3 (1), 56–61.

Swanson, David, Cho, Moon Jung and Eltinge, John: Detecting Possibly Fraudulent Data or Error-Prone Survey Data Using Benford's Law, in: Proceedings of the American Statistical Association (Survey Research Methods Section), 2003, 4172–4177.

Tödter, Karl-Heinz: Das Benford Gesetz und die Anfangsziffern von Aktienkursen. WiSt, 2007, 36 (2), 93–97.

Turner, Charles F., Gribbe, James N., Al-Tayyip, Alia A. and Chromy, James R.: Falsification in Epidemiologic Surveys: Detection and Remediation (Prepublication Draft), Technical Papers on Health and Behavior Measurement, No. 53. Washington DC: Research Triangle Institute, 2002.

Wang, Yongyi and Pedlow, Steven: Detecting Falsified Cases in SCF 2004 Using Benford's Law, in: Proceedings of the American Statistical Association (Survey Research Methods Section), 2005, 3652–3657.

Wetzel, Angela-Jo: Assessing the Effect of Different Instrument Modes on Reinterview Results from the Consumer Expenditure Quarterly Interview Survey, in: Proceedings of the American Statistical Association (Survey Research Methods Section), 2003, 4508–4513.

A Method for Ex-Post Identification of Falsifications in Survey Data

Natalja Menold, Peter Winker, Nina Storfinger, Christoph J. Kemper

Abstract

Falsified survey data can have a substantial impact on data quality. A method is presented here which enables identification of falsifications in survey data. This method is based on a detailed analysis of the motivation behind falsifications, which permits us to derive hypotheses about the properties of data generated by falsifiers. Using these hypotheses indicators of specific statistical properties of falsified interviews are constructed. Using false and real data generated in an experimental setting, we examine which indicators are valuable in terms of their capability to distinguish between falsifiers and non-falsifiers. Classification of interviewers is based on a multivariate analysis. The results are discussed in relation to methodological issues which arise when this approach is applied in real surveys.

Keywords: Interviewer falsifications; Indicators for falsifications; Classification

Acknowledgements

We would like to thank the DFG for its financial support in projects WI 2024/3-1 and ME 3538/2-1 within SPP 1292. We would also like to thank Michael Blohm, Sebastian Bredl, Gesine Güllner, Marie Kristin Opper, Rolf Porst and Viktoria Trofimow for their helpful support.

Identification of Falsifications in Survey Data

Interviewers may be faced with incentives to deviate from prescribed routines for several reasons. These include difficulties in making contact with the target person, complex questionnaires, or a low payment level based only on the number of completed interviews. Moreover, deviations can take place in a number of different ways. Rather subtile forms include surveying a member of a household other than the intended one, or conducting the survey by telephone when face-to-face interviews are required (Groves et al. 2004). Complete or partial fabrication of interviews can be considered to be the most severe form of deviation or falsification.[1] Whilst research on falsifications of interviews is rare in literature,[2]

---

1 The act of fabricating entire interviews is called "curbstoning" by the US Bureau of the Census.

2 It is more common to remove suspicious data prior to further analysis without reporting in detail on the methods used for identifying interviewers at risk or the share of data removed. Notable exceptions are the NKPS, for which the codebook (http://www.nkps.nl/CodeBook/CodeBookFrame.htm) reports in section 11.2 on the procedure and detected

anecdotal evidence indicates a substantial prevalence of the problem, with estimates typically ranging between 1 and 9 %. However, in certain contexts falsification rates may be even higher, for example as reported by Bredl et al. (2012) for a small survey in a non-OECD country.

Fabricated interviews can have serious consequences for statistical analyses based on survey data. Schnell (1991) and Schräpler and Wagner (2003) provide evidence that the effect on univariate statistics may not be so severe. But even a small proportion of fabricated interviews can be sufficient to cause significant biases in multivariate statistics. Schräpler and Wagner (2003) analysed data from the German Socio Economic Panel (SOEP) and found that the inclusion of fabricated data in a multivariate regression reduces the estimated effect of years of education on log gross wages by approximately 80%, even though the share of fabricated interviews was less than 2.5%. Consequently, methods aimed at preventing and/or detecting falsified interviews are of crucial importance in face-to-face survey research.

The most effective way to identify falsified interviews is the re-interview (Biemer and Stokes 1989). However, for reasons of expense it is impossible to re-interview all respondents in large surveys. Therefore, we must ask how the re-interview sample can be optimized to best detect dishonest interviewers. Generally speaking it seems practical to re-interview respondents who have already been interviewed by an interviewer who appears likely to be a falsifier. Hood and Bushery (1997) use the term "at risk" interviewer in such a context. The problem of determining which interviewers are "at risk" was already addressed in the 1980s, although literature on this issue is still scarce. Previous research was aimed at interviewer characteristics signalling "at risk" interviewers. For example, interviewers with a relatively short length of service have been found to be more likely to falsify interviews (Biemer and Stokes 1989; Koch 1995; Schreiner et al. 1988; Wetzel 2003).

Another approach is to check the plausibility of the data collected by interviewers. Hood and Bushery (1997) used several indicators to identify "at risk" interviewers in the National Health Interview Survey (NHIS). For example, they calculated the rate of households labelled ineligible or the rate of households without telephone numbers per interviewer and compared the rates to census data. Koch (1995) used deviancies in data relating to gender and age of interviewed persons, which were known to the surveyor from official personal registers.

---

cases, and the German General Social Survey (ALLBUS), for which identification methods and the number of suspicious cases are reported as well.

Additionally, systematic differences between real and falsified data can be used as criteria to identify "at risk" interviewers. We will refer to such criteria as "indicators" here. Thus, falsifiers may produce distributions of opinions and behaviour which differ from those of real respondents. We refer to such indicators as "content-related". Reuband (1990) found that falsifiers produce more optimistic predictions for future individual economic situations than real respondents. In addition, differences in response behaviour can also be used as indicators. Bredl et al. (2012) found that falsifiers tend to avoid the category "others, please specify" when answering semi-open-ended questions. Schäfer et al. (2005) showed that falsifiers produced less item non-response and less variability in their data. We refer to such indicators, which are not based on differences in substantial responses to survey questions but rather on differences in response behaviour, as "formal" indicators.

The multivariate method proposed by Bredl et al. (2012) simultaneously uses information related to several formal indicators to detect "at risk" interviewers. This method uses only information contained in the data and no other external sources. In doing so, the information from different formal indicators and their dependency structure are used to classify interviewers, for example by means of cluster analyses. This is expected to result in higher selectivity in terms of identifying interviewers "at risk".

The goal of the current study is to build on the method suggested by Bredl et al. (2012) to distinguish between real and falsified data. We aim to do this by introducing additional indicators and improving methods for multivariate data analysis. This research will provide new methods for identifying data falsifications and is intended to complement other methods aimed at quality control of interviewers' work (e.g. re-interviews). Our analysis is based on experimental data in which the correct assignment of real and falsified interviews is known. This allows for evaluation of the outcome of the procedure. The dataset is also substantially larger than that studied by Bredl et al. (2012), resulting in greater power of statistical data analysis.

In the next section we will describe the different indicators. The differences between real and falsified data in terms of these indicators are based on the psychology of survey response and approaches to the motivation of falsifiers. Subsequently, we will provide a description of our procedure. Next, we will present the results concerning the ability of indicators to differentiate between falsified and real data. Finally, we will present conclusions based on our results.

Indicators Sensitive to Falsifications

To identify falsifications we do not rely on individual characteristics of interviewers, but rather on indicators derived solely from the survey data. These indicators should help to distinguish between falsifiers and non-falsifiers. In this section we present the theoretical background which helps to provide an explanation for differences between real and falsified data that have been used as sources of indicators in previous research (e.g. Bredl et al. 2012; Hood and Bushery 1997; Shaeffer et al. 2005). Furthermore, these theoretical approaches are helpful in identifying additional content-related or formal indicators which have not yet been used to detect falsified data.

In terms of content-related indicators we consider differences in information availability between real respondents and falsifiers. In particular, when producing substantial responses to survey questions falsifiers tend to make use of stereotypes related to potential respondents (Hippler 1979; Schnell 1991) since little detailed individual information about the respondents is available to them. Stereotypes are theories about characteristics of a social group shared by another social group (Stroebe et al. 1992). When using stereotypes individuals overestimate differences between social groups as well as similarities within social groups (Tajfel 1969). The next idea in terms of content-related indicators is to deliberately exploit the different levels of information availability between real respondents and falsifiers. Here a special kind of control question comes into play. For example, one can ask "What magazines do you read?" and show a card presenting a list of magazines. Considering that half of the magazines on the list do not actually exist, falsified data would deviate from real data (Noelle-Neumann 2003). In this case it is less likely that real respondents would chose fictitious magazines because exact information is available to them and they only have to look for the magazines in the list they actually. Falsifiers may show a higher probability of choosing a non-existent magazine since they do not expect that some response alternatives are implausible. We assume that indicators based on other fictitious response options produce similar differences between real and falsified data.

When considering formal indicators we rely on the "satisficing model" introduced by Krosnick and Alwin (1987). The authors define satisficing as a superficial kind of information processing in which respondents minimize their cognitive effort when responding to survey questions. Here respondents do not try to find an optimal answer, but rather just an acceptable one. For example, satisficing results in frequently selecting the "do not know" category, in often providing extreme or neutral positions, or – in extreme cases – in providing the same response to every item (straight lining). We assume that real respondents and falsifiers differ in their satisficing behaviour, thus producing differences in formal indicators.

In most cases the predominant motivation of falsifiers is to save time and effort, which consequently results in a higher level of satisficing when compared to real respondents. This can explain falsifiers' avoidance of the "others, please specify" category in semi-open-ended questions, which has been found by previous research (Bredl et al. 2012). The "others, please specify" category was used in questions with a list of nominal response options. A response which does not match with any other available option should fit into the "others, please specify" category. Since the "other" information is known to the real respondent and is easy to provide, this person may have a low tendency to exhibit satisficing behaviour in this setting. In contrast, in the case of falsifiers providing additional information when choosing the category "others, please specify" is associated with a higher level of cognitive effort, since falsifiers then have to create this additional information. In avoiding this high level of cognitive effort falsifiers show a higher level of satisficing behaviour by simply using one of the alternatives presented in the list. Similar behaviour (e.g. higher levels of satisficing) has been found when using filter questions (Hood and Bushery 1997). Filter questions allow a part of the questionnaire to be skipped. Thus, falsifiers use this option to reduce their cognitive effort and to save time.

However, falsifiers try to save time and reduce their cognitive effort only if legitimate opportunities are provided in the questionnaire. Otherwise falsifiers might show less satisficing than real respondents. This tendency is caused by the motivation of falsifiers to avoid detection. This is in line with findings indicating that falsifiers produce a lower level of item non-response than real respondents (Bredl et al. 2012; Shaeffer et al. 2005). Less item non-response on the part of falsifiers has been found for both open-ended and closed-ended questions. Since falsifiers tend to use only legitimate options in reducing their effort, at the same time they naturally avoid less legitimate options (e.g. item non-response). Falsified responses to open-ended questions require a level of effort which is probably still considered to be acceptable since open-ended questions are used quite rarely in standardized surveys. More frequent responding to open-ended questions seems to be inconsistent with less frequent usage of the "others" category in semi-open-ended questions. However, this may not be the case since falsifiers do not produce a high level of item non-response in the case of semi-open-ended questions, but rather a substantial response when they avoid "others, please specify" by ticking one of the available options.

A lower level of satisficing can also explain the previous finding that falsifiers produce less extreme responses than non-falsifiers (Shaeffer et al. 2005). Less extreme responding by falsifiers may also be associated with lower variances, as this has been found in falsified data by previous research (Shaeffer et al. 2005) – mainly for multi-item sets with rating scales. At first glance, lower variances (we refer to this as "non-differentiation") are not in line with the assumption of lower satisficing on the part of falsifiers. However, if we consider that usage of less extreme categories produces lower variances, we can understand higher non-differentiation as a secondary effect of less satisficing when providing less extreme responses. In addition, when using stereotypes falsifiers are not able to produce the same amount of variability in the data as that which is obtained in the real data. This can explain high non-differentiation in falsified data as well.

If we consider other response tendencies which are known for real respondents (Tourangeau et al. 2000) – middle responding, acquiescence, rounding, primacy and recency effects – we are able to generate additional hypotheses about formal indicators. In terms of all of these response tendencies we also assume less satisficing by falsifiers than by real respondents. In the following points we provide definitions and explanations for each of these response tendencies.

- Middle responding refers to the frequent usage of the middle category in rating scales; in doing so moderate opinions or neutral positions are indicated. We assume that falsifiers avoid middle responding so as to not make themselves conspicuous by producing only neutral opinions.
- Acquiescence is a tendency to agree, whereby real respondents tend to provide "Yes/Agree" responses regardless of item content (Messick 1976). This often occurs in multi-item sets with reversed wording of items. Here, respondents provide answers for negatively worded items in the same way that they answer positively worded items. In accordance with our assumption of less satisficing we suppose that falsifiers avoid this response pattern.
- Rounding occurs because respondents tend to answer open-ended questions about numerical information, for instance frequencies or quantities, using round numbers. Round numbers are multiples of 5, 10 or 7, the latter being used with questions related to the calendar (Tourangeau et al. 1997). Here a lower level of satisficing means that falsifiers answer metric questions with round numbers less frequently than real respondents.
- Primacy and recency effects are related to the order of presentation of response options (Tourangeau et al. 2000). These occur if sequences of response categories are presented to respondents, for example with "check all that apply" questions. A primacy effect might appear when response options are presented visually. In this case respondents prefer options located at the beginning of a list, meaning that they tend to choose the first option which seems satisfac-

tory, and they ignore the remaining options. This behaviour is a typical example of satisficing (Krosnick and Alwin 1987). Based on our assumption of a lower level of satisicing on the part of falsifiers, falsifers might produce a lower primacy effect.

- However, in the case of acoustic presentation of response alternatives, rather than primacy being an issue, there is instead a recency effect in real interviews. Recency effect means that respondents show a preference for choosing the last categories mentioned (Krosnick and Alwin 1987). Acoustic presentation is associated with the limited capacity of short-term memory, thus causing a recency effect. A falsifier uses a visual presentation of response categories (e.g. in a CAPI program), and thus a recency effect is less likely to influence falsified data. When generating indicators for primacy and recency effects it is important to use different presentation modes for lists of response categories.

In summary, we expect that falsifiers display reduced effort (higher satisficing) resulting in them choosing the "other, please specify" category less frequently and in them skipping subsequent questions in filter questions more frequently. In contrast, we expect less satisificing related to the remaining formal indicators, meaning that there is more effort on the part of falsifiers as compared to real respondents. This is expected to result in a lower level of item non-response, less frequent extreme and middle responding, acquiescence, rounding, and lower primacy and recency effects on the part of falsifiers. Next, we expect to find higher non-differentiation in falsified data resulting from avoidance of extreme categories and stereotyping on the part of falsifiers. In terms of content-related indicators in particular we suppose that questions with fictitious response categories are powerful in differentiating between real and falsified data. However, in the present study we used only relatively few content-related indicators since in our previous studies we found that formal indicators were associated with higher effects and more stable results than content-related indicators.

We have already tested some of the assumptions described in two preliminary studies (Menold, Storfinger and Winker 2011; Menold and Kemper 2011; Storfinger and Winker 2011). Both studies used subsets of data from the German General Social Survey (ALLBUS) 2008. Along with these subsets of real ALLBUS data, false data were produced by professional interviewers, students and researchers who played the role of falsifiers. These "falsifiers" were provided with some regional and socio-demographic information about real respondents. To test our assumptions concerning different indicators we then compared the real ALLBUS data and the data produced by the "falsifiers". In doing so we used an experimental design first applied by Hippler (1979). The results of these studies showed that "falsifiers" produced less extreme responses, less rounding, a lower recency effect and higher non-differentiation. Also, falsifiers used filter questions to skip some questions more often than real respondents did. In terms of responses to questions about behaviour we found that past political participation was considerably underestimated by falsifiers.

The aim of the current study is to validate the results of these studies and to test additional indicators (described above) which we were unable to test in the preliminary studies. In doing so we identify numerous indicators which we expect to be sensitive to falsifications, and we use them to improve the multivariate method proposed by Bredl et al. (2012).

Test of Indicators

Sample and Procedure

We used the approach by Hippler (1979), described above, for our explorative studies. The study was conducted in the summer of 2011. In order to compare real and falsified survey data as the first step we gathered $N = 710$ real face-to-face interviews. $N = 78$ students from the University of Giessen (age $Md = 24$ years, $IQR = 3$; 59.2% female, 97.9% single) were recruited as interviewers. The real respondents (age $Md = 24$ years, $IQR = 3$; 60% females, 97.6% single) were also students from the same university and they were recruited by the interviewers. On average each interviewer conducted 9.1 ($SD = 1$) interviews, mostly at respondents' homes (58.1%) or on the university campus (28.4%). Each interview took about 30 minutes and each was audio recorded and checked for validity after the study was completed to ensure that interviews were actually conducted and not falsified. Interviewers received a payment of about eight Euros per interview. Interviewers were also provided with 20 Euros to serve as incentives which they could pay to respondents.

The second step was to have interviewers, who had just collected real data, fabricate survey data in the lab. In the process each interviewer received personal descriptions of real survey participants interviewed by his/her colleagues. These descriptions were randomly assigned to interviewers and they contained characteristics that a potential falsifier could obtain through a contact or short interview: e.g. sex, age, subject of study, number of semesters enrolled, marital status, residence, and country of origin. The interviewers now playing the role of falsifiers were briefed on the purpose of the study, which was to improve methods of identifying falsifications of survey data. They were also briefed on their role as falsifiers and on the procedure. The "falsifiers" were instructed to imagine conducting a face-to-face interview with the person described and then to fill in the survey questionnaire as they thought the person described would have responded to it. In this way they should try to provide answers that map the answers of the person described as closely as possible. The interviewers were paid three Euros for each falsified interview. We also provided a prize of 100 Euros for each of the three "best falsifiers". An interviewer could win one of these prizes if their false data remained undetected following application of the multivariate detection method. In this manner we obtained a falsified data set of N = 710 falsified interviews corresponding to the N = 710 real interviews.

Questionnaire

The questionnaire used for collecting both real and false data contained 62 questions which were selected in such a way that they allowed for the construction of indicators sensitive to falsification. We predominantly used questions from the ALLBUS 2008. These questions covered different domains, such as attitudes toward political issues, attitudes toward women's labour force participation, the economic situation, social justice, and political participation. Additionally, we used a questionnaire on personality (Big Five Inventory-10 (BFI-10; Rammstedt and John 2007) used in the International Social Program, ISSP). We also included a question related to magazine reading containing a portion of fictitious magazines (see above) and a measure of knowledge (Vocabulary and Overclaiming Test, VOCT; Ziegler et al. 2012). The VOCT includes terms which might be considered to be of common usage, as well as fictitious terms. The respondents were asked to assess which terms were known to them.

Construction of Indicators

Indicators were constructed based on some specific individual questions, as well as the multi-item measurements of attitude, behaviour, and personality mentioned above. In the case of multi-item measurements respondents indicated their responses on rating scales containing four to seven categories. Response scales were either unipolar or bipolar. In the current study we tested the following indicators:

1) SEMI-OPEN: relative frequencies of choosing the "others, please specify" category in four semi-open-ended questions included in the questionnaire (related to party vote, income sources, intended study degree and kind of study entrance diploma).
2) OPEN: in the questionnaire we used four open-ended questions related to an understanding of left and right orientation, magazines read and participation in sports. We counted the relative frequencies of providing responses to these open-ended questions to give us an indicator value.
3) FILTER: for this indicator we used ALLBUS questions pertaining to the father's education and occupation when the respondents were 15 years old. These questions were asked in a subsequent step of a filter question. Here, choosing one of several response categories (e.g. father not known) allowed respondents to skip subsequent questions about their fathers. The frequency of choosing these categories was used as indicator.
4) INR (Item Non-Response): we counted the frequencies of item non-response across all questions (items) in the questionnaire (except for questions about the respondents' fathers and open-ended questions).
5) In terms of ERS (Extreme Responding Style) we counted the relative frequencies of choosing the most extreme responses on the rating scales, for example "1" or "5" on a five-point rating scale, across all items in multi-item measures (overall 42 items were considered).
6) For MRS (Middle Responding Style) we used only measures with an uneven number of response categories and we counted relative frequencies of choosing the middle category, again in the multi-item measures (in 26 variables).
7) ARS (Acquiescent Responding Style) was estimated based on BFI-10 responses (10 items). The BFI-10 contains five pairs of items which share a substantial amount of content but differ in terms of positive and negative item wording. ARS represents relative frequencies of agreement responses regardless of item direction.
8) ND (Non-Differentiation): we calculated the average standard deviation of responses across all items in multi-item sets (we used six multi-item sets).
9) ROUNDING: we calculated the relative frequency of rounded responses to numerical open questions. These included questions about the number of minutes spent watching television (rounded numbers here were: 30, 60, 90, 120, etc.), body mass information and income/payment information. For the latter students' income from different sources was queried: from parents, from study fundings, income from their own jobs.
10) PRIMACY: we counted how often (relative frequencies) the first two categories were chosen in a list of nominal response categories provided in four questions.

11) RECENCY: for two questions (about social class and social equity) interviewers read five to six categories in a rating scale to respondents. We then calculated how often respondents chose the last category.
12) NEWS: for this indicator we asked respondents what magazines they read and we provided them with a list of them, half of which were fictitious, as response categories. To create an indicator we calculated how often fictitious magazines were chosen (relative frequencies).
13) VOCT: We used the VOCT test and calculated the relative frequency of real words being correctly recognized as terms which actually exist.
14) PARTICIPATION: For this indicator past political activities were asked about using a "check all that apply" question. The relative frequency of political activities checked represents the value for this indicator.

Statistical Analysis

In order to test whether the indicators are sensitive to falsification as a first step we conducted a between-subjects multivariate analysis of variance (MANOVA) for false vs. real interviews. All of the indicators listed in the previous section were used as dependent variables. The MANOVA was conducted using SPSS 20. The values for dependent variables, with the exception of "non-differentiation" (ND), were not normally distributed in either of the treatment groups. However, in the case of large samples (n > 50) with equal group sizes, MANOVA results are quite robust to multivariate non-normality (see for instance Field 2009). Consequently, non-normality should not impact the results of MANOVA in our study since we used N = 710 for each of the treatment groups. The Box-Test pertaining to equality of covariance matrices can also be disregarded if equal group sizes are used (Field 2009). The Levene test only revealed equality of variances in both treatment groups for a few dependent variables. However, Hartley's FMax test showed that the assumption of homogeneous variances within the groups is supported for all dependent variables.

Group membership for false and real data is known within the MANOVA. An alternative analysis is needed in order to find out how useful the indicators might be in detecting false data when group membership is not known. To this end we combined indicators which significantly differed between real and false interviews, as found by MANOVA, by means of multivariate statistical methods. A cluster analysis appears to be a promising way of using the multivariate structure of the indicator values to distinguish between the group of potential falsifiers and the honest interviewers (Bredl et al. 2012).

The central idea of a cluster analysis is to group similar elements together, while elements from different groups are naturally relatively heterogeneous. Due to the fact that we aimed to identify cheating interviewers and not simply single falsified interviews, the ideal outcome of the cluster analysis is the grouping of interviewers into two groups: one containing the falsifiers and the other containing the honest interviewers. Obviously based on a finite number of interviews per interviewer and many other factors affecting the distribution of indicator values such a perfect grouping cannot be expected in real applications. Nevertheless, this method did turn out to identify at least one cluster of susceptible interviewers in past survey applications, and this cluster included most if not all of the actual falsifiers.

To implement the clustering method each interviewer is characterized by a vector of *d* numerical values, whereby *d* corresponds to the number of indicators considered for the analysis. A single value is obtained for each indicator by pooling together all of the interviews done by one interviewer then calculating the values for the indicator based on this set of interviews. The aim of this cluster procedure was also to construct two clusters; one containing falsifiers and the other containing honest interviewers. Apart from the hierarchical clustering (Ward's method) used for the present analysis, we also tested the k-means approach. Since we obtained more reliable results using the hierarchical method as compared to k-means method, we only report the results based on Ward's method here.

In addition to standard hierarchical clustering we also tested a clustering procedure which constructs clusters by global optimization of an objective function. The threshold acceptance (TA) algorithm (Winker 2001) was the method used within this framework to approximate the optimum solution. Starting with a randomly drawn assignment of the elements the TA procedure then assigns one element, also randomly drawn, of one cluster to the other cluster and accepts this new assignment as long as the modified cluster structure exhibits an improved value of the objective function or if it is at least not worse than the previous solution by more than a specific threshold. With this approach we allow for either an increase or a decrease in the quality of cluster structures in order to find the global optimum for the objective function under consideration, or at least a close approximation, after conducting a large number of local search steps. Based on heterogeneity within the clusters, measured by the sum of the pair wise distances in the respective cluster, the objective function used here should be minimized.

Obviously, given the assumptions regarding interviewer behaviour an optimal outcome for the cluster process is the formation of two clusters which can be separated exactly in terms of the indicators. To obtain such a result all interviewers in the “falsifier cluster” should, for example, show lower values for the proportion of “extreme” answers and for choosing the option “others”. This means that falsifiers should be separated from real respondents in exactly the manner which we supposed above. Given that such a perfect clustering is unlikely to occur in a real setting, an ex-post evaluation of the procedure is necessary to validate the approach. This is feasible in an experimental setting such as the one presented in this paper. In order to assess the performance of the method we can consider the proportion of interviewers who are correctly assigned as well as the proportion who are incorrectly classified. When doing this two types of errors can occur: type I errors, which involve a failure to assign a falsifier to the "falsifier cluster", and type II errors, which are "false alarms" indicating erroneously that an interviewer who performed all of his/her interviews correctly produced a falsification.

Results

Differences Between Real and Falsified Data

In this section we present the results regarding differences between real and falsified interviews. We calculated the MANOVAs on an individual case level and on an aggregated interviewer level. In this section we report the results calculated for each individual case since these do not differ from the results on an interviewer level. The results for the entire multivariate model show that there is at least one substantial difference between real and falsified data ($F_{(14,1405)} = 27.95$, $p < .001$, $\eta_p^2 = 0.22$). Next, we consider the univariate results concerning each indicator included in the model (Table 2.1).

Table 2.1: Univariate differences between real and falsified data regarding indicators of cheating

Indicator	Hypothesis	Con-firmed	Falsifica-tions *M (SD)*	Real Data *M (SD)*	$F_{(1,1418)}$	η_p^2
SEMI-OPEN	False < Real	yes	0.02 (0.07)	0.05 (0.11)	28.33***	0.02
FILTER	False > Real	yes	0.17 (0.38)	0.07 (0.27)	27.23***	0.02
OPEN	False > Real	no	0.94 (0.15)	0.97 (0.10)	21.79***	0.02
INR	False < Real	no	2.32 (2.65)	2.44 (2.21)	0.82	0.00
ERS	False < Real	yes	0.21 (0.13)	0.23 (0.11)	5.98*	0.004
MRS	False < Real	no	0.24 (0.12)	0.22 (0.10)	9.58**	0.007
ARS	False < Real	yes	0.46 (0.12)	0.51 (0.13)	65.48***	0.04
ND	False ≥ Real	yes	0.91 (0.18)	0.90 (0.16)	71.21***	0.05
ROUNDING	False < Real	no	0.56 (0.25)	0.45 (0.22)	88.72***	0.06
PRIMACY	False < Real	yes	0.33 (0.52)	0.43 (0.57)	12.21***	0.01
RECENCY	False < Real	yes	0.03 (0.13)	0.06 (0.16)	10.51**	0.007
NEWS	False > Real	yes	0.02 (0.07)	0.01 (0.04)	17.44***	0.01
VOCT	False ≤ Real	yes	0.34 (0.34)	0.44 (0.30)	32.92***	0.02
PARTICIPA-TION	False < Real	yes	0.37 (0.17)	0.43 (0.16)	51.94***	0.04

Note. *** $p < .001$; ** $p < .01$; * $p < .05$. Data for n = 710 real and n =710 falsified interviews. For explanation of indicators see "Construction of Indicators".

As described above we expected to see a limited effort on the part of falsifiers as compared with real respondents, in particular through less frequent usage of the category "others" in semi-open-ended questions and in more frequent responses to filter questions which allow skipping of subsequent questions. As can be seen in Table 2.1, there are significant differences with respect to semi-open-ended and filter questions, and these are in line with our expectations.

For another set of indicators we expected more careful response behaviour – in other words less satisficing – by falsifiers than by real respondents. These indicators were: frequency of answering open-ended questions (OPEN), item non-response (INR), extreme responding (ERS), middle responding (MRS), acquiescent responding (ARS), non-differentiation (ND), rounding, primacy and recency effects. The results support a majority of our hypotheses as they show less extreme responding, less acquiescence, higher non-differentiation, as well as lower primacy and recency effects for falsified data as compared with real data. All of these differences are significant. The results for item non-response (INR) showed fewer unanswered items on the part of the falsifiers, although this result is not significant.

However, there are three significant differences which are not in line with our assumption of less satisficing on the part of falsifiers. In fact, falsifiers demonstrated a higher level of satisficing than real respondents since they provided fewer responses to open-ended questions, more rounded answers to openly asked numerical questions (indicator ROUNDING) and a higher level of middle responding style (MRS).

Next, we confirmed our assumptions concerning a higher selection of fictitious response categories by falsifiers. In fact, falsifiers chose fictitious magazines more often to describe the reading behaviour of respondents (NEWS). Similarly, falsifiers also used fictitious terms more often to describe the vocabulary known to respondents (VOCT). Furthermore, past political participation was strongly underestimated by falsifiers, as has also been shown by our previous explorative research.

In summary, we found that all indicators produced significant differences between real and false data, with the exception of item non-response. Thus, these can be used as indicators of cheating behaviour in the subsequent cluster analysis. The results also show that the assumption about "less satisficing" applied to the majority of respective indicators. However, in the case of three indicators reduced effort by falsifiers was observed. We assume that in this case falsifiers' motivation to remain undetected (which we expected to result in less satisficing) conflicts with their motivation to save time. This is likely since in our experimental design the consequences related to being detected were not nearly as serious as those in a real survey context. In addition we explain the result related to open-ended questions by an excessively high level of effort required from falsifiers in the current study: we included not just one but rather a number of open-ended questions which had to be answered for approximately 10 respondents. To explain the result concerning ROUNDING we compared the current results with results from our previous explorative studies. In the current study we asked for specific individual income information and mainly on this basis we calculated a value for the rounding indicator. In previous studies (e.g. Menold et al. 2011) we used a classic total net household income question which has been commonly used in surveys. This question requires consideration of each person living in the household and all income sources, which may be associated with memory gaps on the part of real respondents, who then round numbers more often than falsifiers do. Therefore, we conclude that availability of exact information on the part of real respondents could explain contradictory results found in our different studies. If exact information is not available in the memory of real respondents (e.g. when asking about household net income) rounding should become more relevant in real data than in falsified data. In contrast, if exact numerical information is known to respondents (e.g. individual incomes) then non-falsifiers will demonstrate less rounding than falsifiers. A future subject of research will be to study in more detail the impacts of the likelihood of detection, task difficulty and information availability on falsifiers' behaviour.

Cluster Analysis for Separating Real and False Data

Based on the significant results of the univariate analysis in terms of differences in indicator values between falsified and real data (see Table 2.1) we decided to consider 13 indicators for the cluster analysis at the interviewer level. The non-response indicator (INR) was omitted due to insignificant results in the univariate analysis (see Table 2.1).

As a first step, we provide the results regarding the cluster performance of each indicator separately. In the second step, we report the results for the multivariate cluster analysis including all 13 indicators simultaneously. The effectiveness of the analysis is always judged based on the proportion of correctly assigned falsifiers as well as the proportion of correctly assigned honest interviewers (non-falsifiers).

As mentioned above we aimed to identify falsifiers and not single falsified interviews. Consequently, this analysis required calculation of the 13 indicator values on an interviewer level. Hence, we had to bring together the indicator values by pooling together all interviews conducted by one interviewer and then we had to calculate the respective indicator value. This was repeated for each of the 156 interviewers (78 falsifiers and 78 non-falsifiers) in the dataset. As reported in Table 2.1 calculating indicator values at the interviewer level produces similar mean values and significant differences between falsifiers and non-falsifiers. But we can also note that the variance of all indicator values is higher within the cheating group than within the group of non-falsifiers.

Table 2.2 shows the results of the hierarchical clustering method. The proportion of falsifiers who were correctly identified ranges between 17% and 91%, while the proportion of correctly classified non-falsifiers varies between 26% and 100%.

Table 2.2: Cluster results based on the hierarchical clustering method (Data: 78 falsifiers and 78 non-falsifiers)

	CLUSTER A (Falsifiers)			**CLUSTER B (Non-Falsifiers)**		
Indicator	**Cluster-size**	**Falsifiers** (% of all falsifiers)	**Non-Falsi-fiers**	**Cluster-size**	**Non-Falsifiers** (% of all non-falsifiers)	**Falsi-fiers**
SEMI-OPEN	129	71 (91%)	58	27	20 (26%)	7
FILTER	43	32 (41%)	11	113	67 (86%)	46
OPEN	39	29 (37%)	10	117	68 (87%)	49
ERS	33	29 (37%)	4	123	74 (95%)	49
MRS	102	51 (65%)	51	54	27 (35%)	27
ARS	83	57 (73%)	26	73	52 (67%)	21
ROUNDING	21	21 (27%)	0	135	78 (100%)	57
PRIMACY	55	36 (46%)	19	101	59 (76%)	42
ND	86	51 (65%)	35	70	43 (55%)	27
RECENCY	127	70 (90%)	57	29	21 (27%)	8
NEWS	15	13 (17%)	2	141	76 (97%)	65
VOCT	31	29 (37%)	2	125	76 (97%)	49
PARTICIP.	35	27 (35%)	8	121	70 (90%)	51
ALL 13 INDICAT.	**44**	**43 (55%)**	**1**	**112**	**77 (99%)**	**35**
Subset 1	23	23 (29%)	0	133	78 (100%)	55
Subset 2	145	74 (95%)	71	11	7 (0.1%)	4
Subset 3	**79**	**57 (73%)**	**22**	**77**	**56 (72%)**	**21**

When comparing results for the 13 indicators we can see that acquiescence (ARS) and non-differentiation (ND) are indicators which yield outstanding results in terms of identifying real and falsified data. When using acquiescence as an indicator we found that 73% of falsifiers and 67% of non-falsifiers were correctly identified. When using the ND indicator 55% of non-falsifiers and 65% of falsifiers were revealed.

With respect to success only in terms of identifying falsifiers, Table 2.2 shows that nearly all falsifiers could be identified using the RECENCY indicator (90%) as well as the SEMI-OPEN indicator (91%). However, this result is not satisfactory because these two indicators produce the largest "falsifier-clusters" (see Table 2.2). More than two thirds of all interviewers are assigned to this cluster and consequently a high proportion of false alarms is produced. This is also true of the midpoint-ratio (MRS), which reveals more than half of the falsifiers (65%) but the obtained falsifier-cluster itself contains approximately 65% of all interviewers. All other indicators produced fewer false positives but they were only able to identify less than half of the falsifiers when they were considered individually (the proportion varies between 17% and 46%).

While the results related to identifying falsifiers using individual indicators are only satisfying in some cases, identifying non-falsifiers was somewhat more successful. Eight of the 13 indicators detected more than three quarters of all non-falsifiers (proportions range between 76% and 100%) but at the same time high proportions of false negatives are produced. All other indicators were able to identify less than 70% of the non-falsifiers.

As proposed by Bredl et al. (2012), the proportion of correctly assigned interviewers may increase if all 13 indicators are used simultaneously in a multivariate analysis. This way not only the identifying power of each indicator is exploited, but also the dependency structures between indicators, which could increase the overall quality of assignment into the correct group. In fact the results for our dataset are quite satisfying (see Table 2.2) given that assignment to the "non-falsifier-cluster" is successful in almost all cases (99%) and slightly more than half of the falsifiers were assigned to the falsifiers' cluster (55%).

To conclude whether the multivariate approach performs better than the 13 single analyses we must consider the precision of the cluster result, for example by investigating cluster sizes in relation to the proportion of correctly assigned interviewers. From the point of view of practitioners in polling agencies we prefer highly precise values, especially for the "falsifier-cluster". If polling agencies use the cluster result for further checks, such as re-interviews, the number of non-falsifiers in the "falsifier-cluster" should be as small as possible, and of course the number of falsifiers should be as high as possible. Thus, polling agencies could avoid inspecting a high proportion of false alarms. When considering the precise values for the 13 single analyses, we can notice that a few indicators produce very small proportions of false alarms (e.g. ROUNDING and NEWS) but at the same time less than half of the falsifiers are assigned to the respective "falsifier-cluster". Only the multivariate approach provides a satisfactory proportion of correctly assigned falsifiers and at the same time a very small proportion of false alarms. Hence, we recommend relying on the multivariate approach.

Nevertheless, we were interested in increasing the proportion of correctly assigned falsifiers. In doing so we considered smaller subsets of indicators to investigate how the results change. The selection of indicators to be used simultaneously was based on the results obtained from the 13 single (univariate) analyses. Firstly we decided to incorporate only the acquiescence indicator (ARS) and the non-differentiation value (ND) because of their outstanding results in the univariate analysis (Subset 1). Secondly we investigated the SEMI-OPEN and RECENCY indicators because they delivered the highest values in terms of identifying falsifiers in the univariate analyses (Subset 2). For the last subset of indicators we chose all indicators which performed better than the multivariate approach (using all 13 indicators), either in identifying falsifiers (SEMI-OPEN, MRS, ARS, ND and RECENCY) or in identifying non-falsifiers (ROUNDING) (Subset 3).

Table 2.2 shows the results for the three different subsets. The first two subsets succeed only in increasing either the proportion of identified falsifiers or the proportion of correctly assigned non-falsifiers. However the third subset provided high values for both shares, even if the number of false alarms in the "falsifier-cluster" was higher than the value obtained when using all 13 indicators. Consequently, the number of indicators used for the hierarchical cluster analysis depends on the specific interest. When the aim is to produce a precise "falsifier-cluster", for example with low false alarm rates, all 13 indicators should be used. But if we are interested in correctly assigning a high share of all interviewers we might use a smaller subset of indicators.

In addition to hierarchical clustering we also applied a global clustering method based on heuristic optimization (TA) to our data set. In terms of the results for the 13 analyses using only one indicator separately (see Table 2.3) we found that using the global clustering method produces high proportions of correctly assigned falsifiers. Ten of the 13 univariate analyses identified more than half of the 78 falsifiers in the data set (proportions vary between 54% and 83%). Identifying more than half of the non-falsifiers was successful in all 13 single analyses. Also the multivariate approach when using 13 indicators leads to very satisfying results. It was possible to correctly assign 82% of falsifiers and 92% of non-falsifiers.

If we compare these results with those obtained using hierarchical clustering we can notice that the global clustering method provides equal cluster sizes in the majority of the cases. This may be advantageous for our analysis because our dataset contains 50% falsifiers and consequently the proportions of falsifiers revealed are almost always higher than those when using the hierarchical clustering method. Hence, we recommend using the global clustering method instead of the hierarchical approach when the aim is precise differentiation between falsifiers and non-falsifiers. The global clustering method using all 13 indicators simultaneously leads to even higher proportions of correctly assigned interviewers than the hierarchical approach using the small subset of indicators. If we take a look at the results of the TA approach using the three subsets (see Table 2.3) we can notice that the proportions of correctly assigned falsifiers and non-falsifiers are slightly less convincing when compared to the multivariate analysis using all 13 indicators.

Table 2.3: Cluster result based on the global clustering method (TA) (Data: 78 falsifiers and 78 non-falsifiers)

	CLUSTER A (Falsifiers)			CLUSTER B (Non-Falsifiers)		
Indicator	**Cluster-size**	**Falsifiers** (% of all falsifiers)	**Non-Falsifiers**	**Cluster-size**	**Non-Falsifiers** (% of all non-falsifers)	**Falsifiers**
SEMI-OPEN	105	65 (83%)	40	51	38 (49%)	13
FILTER	47	33 (42%)	14	109	64 (82%)	45
OPEN	58	37 (47%)	21	98	57 (73%)	41
ERS	81	43 (55%)	38	75	40 (51%)	35
MRS	72	42 (54%)	30	84	48 (62%)	36
ARS	83	57 (73%)	26	73	52 (67%)	21
ROUNDING	72	55 (71%)	17	84	61 (72%)	23
PRIMACY	82	48 (62%)	34	74	44 (56%)	30
ND	69	46 (59%)	23	87	55 (71%)	32
RECENCY	94	56 (72%)	38	62	40 (51%)	22
NEWS	40	28 (36%)	12	116	66 (85%)	50
VOCT	71	43 (55%)	28	85	50 (64%)	35
PARTICIP.	74	50 (64%)	24	82	54 (69%)	28
ALL 13 INDICAT.	**70**	**64 (82%)**	**6**	**86**	**72 (92%)**	**14**
Subset 1	77	56 (72%)	21	79	57 (73%)	22
Subset 2	97	61 (78%)	36	59	42 (54%)	17
Subset 3	**77**	**58 (74%)**	**19**	**79**	**59 (76%)**	**20**

In summary applying the hierarchical clustering method is inferior to using the global clustering method (TA). Using Ward's method often results in low proportions of false alarms, but this comes at the cost of a high proportion of overlooked falsifiers. Hence, we recommend using the global clustering procedure (TA) in order to achieve a relatively high proportion of revealed falsifiers.

Conclusions

The aim of the study presented in this article was to test numerous indicators in terms of their effectiveness in helping to identify falsified data in face-to-face surveys. The main idea behind this approach was to use differences between real and falsified data as a source of indicators, and to combine these within a multivariate cluster analysis.

The effectiveness of indicators was expected to result from differences in cognitions and motivation between real respondents and falsifiers. We were able to confirm the majority of our assumptions concerning these differences.

In addition, nearly all of the indicators tested are useful for identifying falsified data. Using a selected set of indicators leads to the best performance for the clustering method. This shows that some indicators are more powerful than others. We were able to identify most of the falsifiers in the large data set produced in our experimental setting. However, it was not possible to obtain convincing results with a standard clustering method (e.g. hierarchical cluster). On the other hand an optimised clustering method (TA) was more effective. Further research is needed to verify the extent to which this result can be generalized to other surveys and contexts. In real settings we assume that the proportion of falsifiers is much smaller, ranging perhaps between 5 and 8 % (Schnell 1991). In such a setting identifying falsifications based on the clustering method might be even more successful (Storfinger and Winker 2011). To test this assumption we created synthetic data by means of bootstrapping. This implies that the number of falsifiers is modified artificially by re-sampling individual interviews of randomly selected interviewers. More details on this procedure and the obtained results are provided by Storfinger and Winker (2013) in this volume.

With regard to the number of clusters created in our identification method further developments should be considered. Thus, it could be helpful to create three (instead of two) clusters. In doing so falsifiers themselves could be clustered into two groups: one cluster containing interviewers who falsified data in the way that we have supposed (see hypotheses) and the second cluster containing falsifiers who falsified employ a different strategy. For instance, we found that typical falsifiers avoid extreme answers (ERS), but the dataset shows that there are also some falsifiers who prefer extreme answers even more than non-falsifiers do. First analyses showed that the cluster result improves significantly if we calculate three clusters, but this is only true for a few indicators. Further research on the motivation of falsifiers is necessary in order to understand the relationship between indicator values and the number of clusters.

Despite these remarks regarding possible further developments, it also appears to be important to test the detection method presented in this paper in a real survey setting. This method can be used in surveys as a supplement to other control procedures during data collection. Identifying interviewers who are "at risk" by using this multivariate approach would be more productive for subsequent extended controls compared with single indicators, a method used in current survey practice (e.g. differences in sex and age, as described by Koch 1995). In this way it would be possible to obtain not only higher quality data but also higher validity data regarding the prevalence of falsifications in surveys.

Bibliography

Biemer, Paul. P., Stokes, S. Lynne.: The optimal design of quality control samples to detect interviewer cheating, Journal of Official Statistics, 5(1), 23-39, 1989.

Blasius, Jörg, Thiessen, Victor: Detecting poorly conducted interviews. In Winker, Peter, Menold, Natalja and Porst, Rolf (Eds.): Survey Standardization and Interviewer's Deviations – Impact, Reasons, Detection and Prevention. Peter Lang, Frankfurt a. M., 2013, pp. 67 – 88.

Bredl, Sebastian, Winker, Peter, Kötschau, Kerstin: A statistical approach to detect interviewer falsification of survey data, Survey Methodology, 38(1), 1-10, 2012.

Field, Andy: Discovering statistics using SPSS, Singapore, Sage, 2009.

Groves, Robert M., Fowler, Floyd Jackson Jr., Couper, Mick P., Lepkowski, James M., Singer, Eleanor, Tourangeau, Roger: Survey Methodology, New York, Wiley, 2004.

Güllner, Gesine, Porst, Rolf: Identifikation von Fälschungen in Surveys. Bericht über den kognitiven Test im Rahmen des Projekts IFiS, GESIS-Working Papers, 16, (2012).

Hippler, Hans-Jürgen: Untersuchung zur "Qualität" absichtlich gefälschter Interviews, ZUMA-Arbeitspapier, Mannheim, 1979.

Hood, Catherine C., Bushery, John M.: Getting more bang from the reinterview buck, Identifying "at risk" interviewers, Proceedings of the American Statistical Association, Survey Research Methods Section 27, 820-824, 1997.

Koch, Achim.: Gefälschte Interviews, Ergebnisse der Interviewerkontrolle beim ALLBUS 1994, ZUMA-Nachrichten, 36, 89-105, 1995.

Krosnick, Jon A., Alwin, Duane F.: An evaluation of a cognitive theory of response-order effects in survey measurement, Public Opinion Quarterly, 51, 201-219, 1987.

Menold, Natalja, Kemper, Christoph J.: Survey response characteristics as indicators for detection of falsifications, Paper presented at the 4th Conference of the European Survey Research Association (ESRA), Lausanne, Switzerland, 2011, July.

Menold, Natalja, Storfinger, Nina, Winker, Peter.: Development of a method for ex-post identificationof falsifications in survey data, Proceedings of New Techniques and Technologies for Statistics - NTTS 2011, Brussels, Belgium, 2011.

Messick, Samuel J.: The psychology of acquiescence, An interpretation of research evidence, In: Berg, I. A. (Ed.), Response Set in Personality Assessment, Chicago, Aldine Publishing Company, 1967.

Noelle-Neumann, Elisabeth: Ein Wissenschafter muss Neuland betreten, Wiener Zeitung, April, 17., 2003 (http://www.philosophische-praxis.at/noelle.html, August 30 2012)

Rammstedt, Beatrice, John, Oliver P.: Measuring personality in one minute or less, A 10-item short version of the Big Five Inventory in English and German, Journal of Research in Personality, 41(1), 203-212, 2007.

Reuband, Karl-Heinz: Interviews, die keine sind, "Erfolge" und "Mißerfolge" beim Fälschen von Interviews, Kölner Zeitschrift für Soziologie und Sozialpsychologie, 42(4), 706-733, 1990.

Schäfer, Christin, Schräpler, Jörg-Peter, Müller, Klaus-Robert, Wagner, Gert G.: Automatic identification of faked and fraudulent interviews in the German SOEP, Journal of Applied Social Science (Schmollers Jahrbuch), 125 (1), 183-193, 2005.

Schnell, Rainer: Der Einfluss gefälschter Interviews auf Survey Ergebnisse, Zeitschrift für Soziologie, 20(1), 25-35, 1991.

Schräpler, Jörg-Peter, Wagner, Gert G.: Identification, characteristics and impact of faked interviews in surveys. An analysis by means of genuine fakes in the raw data of SOEP, IZA Discussion Paper Series, 969, 2003.

Schreiner, Irwin, Pennie, Karen, Newbrough, Jennifer: Interviewer falsification in census bureau surveys, Proceedings of the American Statistical Association, Survey Research Methods Section XII, 491-496, 1988.

Shaeffer, Eric M., Krosnick, Jon A., Langer, Gary E., Merkle, Daniel M.: Comparing the quality of data obtained by minimally balanced and fully balanced attitude questions, Public Opinion Quarterly, 69(3), 417-428, 2005.

Storfinger, Nina, Winker, Peter: Robustness of clustering methods for identification of potential falsifications in survey data, ZEU Discussion Paper 57, Giessen, 2011.

Storfinger, Nina, Winker, Peter: Assessing the performance of clustering methods in falsification using bootstrap. In Winker, Peter, Menold, Natalja and Porst, Rolf (Eds.): Survey Standardization and Interviewer's Deviations – Impact, Reasons, Detection and Prevention. Peter Lang, Frankfurt a. M., 2013, pp. 49-65.

Stroebe, Wolfgang, Hewstone, Miles, Codol, Jean-Paul, Stephenson, Geioffrey: Sozialpsycholgie. Eine Einführung. Berlin et al.: Springer, 1992.

Tajfel, Henri. Cognitive aspects of prejustice, Journal of Social Issues, 25, 79-97, 1969.

Tourangeau, Roger, Rasinski, Kenneth A., Jobe, Jared B., Smith, Tom W., Pratt, William F.: Sources of error in a survey on sexual behavior, Journal of Official Statistics, 13(4), 341-365, 1997.

Tourangeau, Roger, Rips, Lance J., Rasinski, Kenneth A.: The psychology of survey response, Cambridge University Press, 2000.

Wetzel, Angela-Jo: Assessing the effect of different instrument modes on reinterview results from the consumer expenditure quarterly interview survey, Proceedings of the American Statistical Association, Survey Research Methods Section 435, 4508-4513, 2003.

Winker, Peter: Optimization heuristics in econometrics, Applications of threshold accepting, Chichester, Wiley, 2001.

Ziegler, Matthias, Kemper, Christoph J., Rammstedt, Beatrice: The vocabulary and overclaiming test (VOC-T), Manuscript submitted for publication, 2012.

Assessing the Performance of Clustering Methods in Falsification Identification using Bootstrap

Nina Storfinger and Peter Winker

Abstract

Data quality in face-to-face interviews might be heavily biased by interviewers' deviant behaviour. Falsifications, as one kind of deviant behaviour, occur if the interviewer fabricates all, or at least a part of the answers in the questionnaire. Regarding the identification of such falsifications we apply a clustering method based on differences in answering patterns of honest and cheating interviewers. Due to the lack of available datasets including identified falsifications, we resort to a bootstrap analysis in order to examine the performance and robustness of this identification method. The bootstrap approach enables us to create further synthetic datasets which differ with regard to the sample size, e.g. the number of interviewers in the dataset, and to examine the sensitivity of the clustering method. The results show that the falsifiers are most often correctly assigned if their share is low and the share of correct assignments slightly increases with a lower total number of interviewers.

Keywords: Interviewer falsifications; cluster analysis; bootstrap method.

Acknowledgements

We gratefully acknowledge financial support through the DFG in projects WI 2024/4-1 and ME 3538/1-1 within SPP 1292.

Introduction

The problem of interviewers' deviant behaviour might occur in every interview situation and therefore is discussed in detail in the existing literature. Bredl et al. (2013, this volume) and Menold et al. (2013, this volume) address the identification of falsifications which occur when the interviewer intentionally deviates from the prescribed rules and fabricates parts of or the whole interview. The current contribution deals with the evaluation of the identification method proposed in Menold et al. (2013, this volume).

The idea of the ex-post identification method was developed by Bredl et al. (2012) and makes use of a cluster analysis which exploits differences in the answering behaviour of correct interviewers and cheaters. By doing this we consider specific characteristics and answer patterns in the interviews and transform them into indicators (see also Menold et al. 2013, this volume). Consequently, the objective of this method is to cluster the interviewers with regard to their indicator values, in two clusters.

The main research question to be answered within this contribution is the assessment of the performance of the method. To this end, it would be necessary to apply the method to a large set of different datasets, which might also differ in some characteristics, e.g. the total number of interviewers or the share of falsifiers. Appropriate data should also contain already identified falsifications because only a priori knowledge about existing falsifiers allows us assessing the performance of the cluster analysis, i.e. the calculation of the share of correct and/ or false assignments. But in the majority of real datasets, the potential falsifications are either removed from the dataset before the data is made available for the public (Schäfer et al. 2005) or have not been identified. This is why we have to resort to a bootstrap based approach.[1] This approach allows us to generate many synthetic datasets while modifying characteristics of the original dataset, like the number of interviewers or the share of cheaters. Thus, we can examine whether the performance of the cluster analysis is sensitive to these modifications and, furthermore, we obtain estimates of the distribution of the results.

The contribution is organized as follows. First, we briefly present the principles of the cluster method used for the identification of the falsifications. The bootstrap method and the specific application to our data will be described in the following section. In the next section we briefly describe the dataset and the indicators used and show the results of the bootstrap method. Finally we conclude and provide a short outlook on future methodological developments and aims regarding the identification of falsifications.

The Clustering Procedure

The procedure for the identification of potential falsifications abstracts from the specific questionnaire content and exploits the finding that cheaters' answering behaviour is different from the data collected by correct interviewers (Menold et al., 2013, this volume). Consequently, the method builds on the idea that it might be possible to separate the two types of interviewers regarding properties of the collected data (Bredl et al. 2012) by means of a cluster analysis.

In an ideal setting all cheaters would be assigned to the same cluster, labelled as "cheater cluster". But as usual in statistical analyses we expect also some misclassification, i.e. two types of errors. The first one occurs when honest interviewers are classified as falsifiers ("false alarms"). The second error refers to a more fatal case, namely the oversight of cheaters. Thus, we have to consider

1 For an introduction to the principles of the bootstrap method the reader is referred to, e.g., Efron (1982) and Chernick (2008). For early applications in the context of cluster analysis see Jain and Moreau (1987) and Peck *et al.* (1989).

besides the correct clustering also the share of "false alarms" and/ or undiscovered falsifiers.[2]

Thinking in more practical terms, both types of errors have different implications for the usage of the identification method in a polling agency. As the method should serve only as a first step of further check-ups, e.g. reinterviews (Schreiner et al. 1988) or postcard checks (Hauck 1969), the share of revealed cheaters might increase when the cases labelled (by our method) as potential falsifiers are recontacted. Thereby, it could be checked if the potential falsifiers are indeed cheaters. However, at the same time this number of interviewers labelled as potential falsifiers should be as small as possible to reduce cost and time of the follow-up checks. If the number of interviewers and interviews to be checked is too high and, consequently, the share of false alarms is very high, the follow-up procedures might be considered as inefficient.

On the other hand, if the number of interviewers and interviews to be checked, as well as the share of false alarms is very small, the check-ups would result in more reliable findings. But at the same time the data analysis might still be biased because of a potentially high number of overlooked falsifiers still present in the dataset. Thus, in this contribution, the results of our identification method are evaluated focusing on the share of revealed cheaters, i.e. a certain amount of false alarms will be accepted.

The huge literature about clustering procedures discusses a wide range of cluster methods. Most famous methods are hierarchical procedures such as Ward's method (1963) or iterative approaches like k-means. Alternatively, one might think about enumerating all possible assignments to two groups and select the one corresponding to the optimal value of a given objective function (Bredl et al. 2012). The latter approach, however, is only feasible for a small number of interviewers and therefore not appropriate for our dataset. We can also implement a heuristic optimization approach to approximate a global optimum (Winker 2001). But the application of this approach to the present dataset (see Menold et al. (2013, in this volume) shows that the results are inferior to the results obtained by the hierarchical cluster method. Hence, for the application presented here, we stick to a hierarchical method, namely Ward's method. In Ward's method the criterion for merging two clusters at any given step is the variance within the clusters. Consequently, out of all pairs of existing clusters

2 Of course the reliable calculation of the share of correct and/ or wrong assignments could only be done if all falsifiers are already revealed, e.g. by applying another identification method. Nevertheless, the cluster analysis is also appropriate to reveal potential falsifiers without having run another identification method. In this situation we have to decide which of the clusters is most likely to contain the cheaters. We explain the principles of this decision in the next sections.

the pair resulting in the lowest increase of the sum of in-cluster variances will be merged.

At the end of the cluster procedure, the obtained clusters have to be labelled meaning that the cluster containing the cheaters should be identified precisely. For this decision we use some a priori assumptions on interviewers' behaviour. If we assume that cheaters tend to produce a smaller share of extreme answers in scale questions, the cluster exhibiting the smaller share of extreme answers could be labelled as the "cheater cluster", when aiming to construct two clusters. Due to our multivariate approach, i.e. including more than one indicator, we might calculate the number of indicators which point in the direction of falsifications. This means that we assume that the clustering does not always result in a perfect grouping and therefore not all a priori assumptions will be matched by the two clusters. The cluster which matches more assumptions regarding falsifications is then labelled as the cluster containing the falsifiers. Besides the number of indicators pointing in the right direction, we can consider the cluster means concerning the indicator values. For this approach we calculate the mean values of the standardized indicators for every cluster and sum them up. If the indicators are defined such that smaller values should point towards potential falsifications, the cluster with the smaller sum might be labelled as "cheater cluster". The advantage of this approach is that it can also be applied for an even number of indicators. By only counting the number of indicators pointing in the right direction, as mentioned before, it might occur that it is not decidable which one is the falsifier cluster if the number of indicators incorporated is even. Thus, we use the approach making use of the sum of standardized indicator values for our analyses.

Finally we have to assess the cluster result. For this evaluation we consider the share of correctly identified interviewers. Alternatively, one might consider only the share of correctly identified cheaters, i.e. to what extent no potential falsification remains unnoticed, or the share of correctly identified honest interviewers, i.e. to what extent "false alarms" have been avoided. We also have a look at these shares.

The Bootstrap Method

The findings of the multivariate cluster analysis in Menold et al. (2013, this volume) show that we are able to identify nearly all of the honest interviewers and a substantial share of falsifiers for the given dataset. Thus, we would like to examine the robustness of the method across datasets with possibly different characteristics. Unfortunately, the access to further datasets with known falsifications is very limited. Hence, we resort to a bootstrap method for constructing artificial datasets. This way we are able to create synthetic data with specific settings for the characteristics we are interested in. For example, datasets might differ in the total number of interviewers, the share of falsifiers in the dataset and the number of conducted interviews per interviewer. In this chapter, we consider modifications of these three characteristics (see also Storfinger and Winker 2011).[3]

Bootstrap as the related Jackknife approach belongs to the class of resampling methods. With this said, the first step of our bootstrap procedure is to draw at random a specific number of interviewers out of the original dataset with replacement where every interviewer has the same probability to be chosen. By doing this we can modify the number of interviewers in the dataset easily. This is also true for the share of falsifiers in the dataset. To this end, we first draw the required number of falsifiers at random with replacement out of the original dataset. Afterwards we apply the same procedure to the subset of honest interviewers.

Once a new synthetic dataset with specific properties is built, we calculate the indicators for each interviewer and conduct the cluster analysis explained above. Finally we evaluate the result of the assignment by means of the calculation of the share of correctly assigned interviewers or the share of misclassifications. Then, we start from the beginning with generating a new artificial dataset. This procedure is repeated many times, typically at least 1000 times. Eventually, i.e., after having computed the indicator(s) of interest for all bootstrap samples, we can report statistical information on the distribution of these indicators such as mean and variance, but also in form of empirical distribution functions, e.g., as histograms in graphical form.

As mentioned above, we use the bootstrap technique to analyze the effect of different modifications of the sample layout on the quality of the results (see Table 3.1). We start with varying the total number of interviewers in the dataset. We are interested in this kind of experiments and the performance of the clustering method when decreasing the number of interviewers because we would like to know if we are able to replicate our previous findings (Storfinger and Winker, 2011). In addition, we are interested if the modification of the total number of interviewers affects the share of correctly assigned falsifiers. We expect that a smaller number of interviewers will make it easier to keep the rate of misclassification of both types low.

3 In our previous analysis (Storfinger and Winker, 2011) we were also interested in the number of questions of a specific type per interview. Due to space restrictions we concentrate here only on the three dimensions mentioned.

The same aspects are considered for the second setting where we modify the number of interviews conducted by every interviewer. The original value of about 10 interviews per interviewer could be sufficient to draw general conclusions but again we would like to see how the cluster results are affected by de- and increasing the number of interviews, respectively. It is expected that a higher number of interviews per interviewer might improve the precision of the estimation of the indicator values and, therefore, improve their discriminatory power.

The last experimental setup refers to the specific share of falsifiers in the dataset. The division of falsified and real interviews in our original experimental dataset (50:50) does not allow to draw general conclusions on the method regarding, primarily its expected performance for a smaller share of falsifiers or the dependence of this performance on specific properties of the dataset. Thus, we are particularly interested whether the method still performs well if the share of falsifiers becomes high (95%) or extremely low given that anecdotal evidence suggests that actual prevalence of falsifications might be of the order of 5% to 10% rather than the 50% set in our experimental dataset. In order to compare the bootstrap result with the findings for the original dataset, we vary only one parameter of the dataset at a time meaning that all other parameters are kept close to their original values. Table 3.1 shows the values used.

Table 3.1: Experimental Design for Bootstrap Runs

Dimension	Original Sample	Values for Bootstrap				Remarks
Number of Interviewers/ Falsifiers	78/78	75/75 142/8	40/40 76/4	25/25 47/3	10/10 19/1	50% Falsifiers 5% Falsifiers
Number of Interviews per Interviewer	10	5	20	50	100	78 Falsifiers/ 156 Interviewers
Number of Falsifiers among Interviewers	78/156	9/147 4/76 1/19	31/125 16/64 4/16	78/78 40/40 10/10	148/8 76/4 19/1	156 Interviewers 80 Interviewers 20 Interviewers

Results

The Dataset and Indicators

The data used for the empirical application have been produced in an experiment within the frame of the research project "Identification of Falsification in Survey Data" (funded by the German Research Foundation) explained in more detail in

Menold et al. (2013, this volume). The survey consisted of 78 interviewers who conducted 710 real interviews (all verified by audio recording) and 710 falsifications, so that the whole dataset comprises 156 interviewers and 1420 interviews. Applying the clustering methods described before to this dataset results in a satisfying assignment (see again Menold et al., 2013, this volume). Using all 13 available indicators and Ward's method for clustering, 43 of the 78 falsifiers were assigned to the cluster labelled "potential falsifications" (55%) and 77 of the correct interviewers were assigned to the "honest-cluster" (99%). Using a subset of seven indicators as presented below changes our result in the way that we assign a higher share of falsifiers correctly (69%) but a smaller share of honest interviewers (88%). However, this result is still satisfying because more than half of the cheaters (54 out of 78) and more than half of the honest interviewers (69 out of 78) are correctly assigned simultaneously.

As described above our identification method requires the measurement of specific characteristics and answer patterns in the questionnaire. Additionally, we aim to identify cheating interviewers and not single interviews. Therefore, the indicators are measured on the interviewer level. Thus, for every interviewer we pool all conducted interviews and calculate the values for the specific indicators. At the end, every interviewer is characterised by n values, when we consider n indicators (see Bredl et al. 2012).

The selection of the specific indicators used for the present analysis is mostly based on the results reported in Menold et al. (2013, this volume). In this complementary work Menold et al. proposed 13 indicators and applied them within the identification method studied here to the same data. Further ideas and theoretical derivations referring to the choice of indicators in this context can be found in the literature (Schnell 1991, Koch 1995, Hood and Bushery 1997, Bredl et al. 2012). Due to the promising results of Menold et al. (2013, this volume) when using a subset which contains only six out of the 13 indicators, we decide to use them. To this subset we add another indicator which is not used in Menold et al. (2013, this volume) but already applied in our previous work (Storfinger and Winker, 2011). In the following we briefly explain these seven indicators as well as the differences of indicator values found between honest and dishonest interviewers.

First, we calculate the share of checked "others" options in all semi-open questions (which offer this option). We noticed (Menold et al., 2013,this volume) that the values for this indicator (SEMI-OPEN) are lower for cheaters than for honest interviewers. The next two indicators refer to scale questions. First we count the number of ticked midpoint answers and expect this value (MRS) to be higher for cheaters. Second, we measure the total range of answers in scale questions (ND). Consequently, we recognized that falsifiers show a smaller range of categories due to their avoidance of extreme answers.

A further specific answer pattern is the so called acquiescence-behaviour. Here, falsifiers tend to agree in opinion questions less often than real respondents in honest interviews. For this indicator (ARS) we calculate the share of compliance in appropriate questions.

Next we concentrate on an indicator referring to the presentation mode of questions. We noticed a recency effect meaning that the respondents choose more often the last read items in a list of items (Menold et al., 2013, this volume). We calculate the share of chosen last items and assume the values of this indicator (RECENCY) to be lower for falsifiers.

Finally, we refer to answers to metric questions which ask for frequencies or amounts. We noticed that falsifiers show more rounded answers than honest interviewers, so we calculate the share of rounded answers (ROUNDING) in all questions which require metric answers. As mentioned in further analyses (Bredl et al. 2012, Storfinger and Winker 2011) we are also interested in an additional indicator based on metric questions. This indicator makes use of the observation that the distribution of the first digits of many metric variables – including monetary measurements – can be well approximated by a specific distribution, the so called Benford's Distribution or Benford's Law (Benford 1938). Given that it appears to be difficult to reproduce this distribution, intentionally or unintentionally, when falsifying questionnaires, it is assumed that the difference between the empirical distribution of first digits and the theoretical distribution (measured by a χ^2-type statistic) is larger for falsified interviews.[4] Hence, we use seven indicators for the analysis presented here. Of course, alternative or additional indicators are conceivable for this type of analysis (see Storfinger and Opper 2011) and are studied in complementary research (Menold, Storfinger and Winker 2011).

Number of Interviewers

We start with the analysis of the first setting, i.e., the variation of the number of interviewers in the dataset. Figures 3.1 and 3.2 summarize the findings. For all settings we consider 150, 80, 50 and 20 interviewers. The share of falsifiers is

[4] As an alternative to this standard setting, we also consider the difference in the distribution of the first digit for one interviewer as compared to the distribution of the first digit of all other interviewers (see Storfinger and Winker 2011). Results for the present data show that falsifiers actually show higher χ^2-values for this type of indicator but the overall result of the multivariate setting does not improve by including this alternative to the Benford indicator.

about 50% for the first setting as in our real dataset, while for the second setting a lower share, namely around 5% is assumed. The latter number corresponds closer to the anecdotal reports about identified falsifications in survey data. Thus, it has been our interest to check how robust the proposed methodology is when the share of falsifiers might be low and – possibly – at the same time the number of interviewers involved might be small. For this experiment, the number of interviews per interviewer is always the same as in the original dataset (about 10 interviews per interviewer). In addition we use here the small subset of seven indicators as mentioned above.

The figures show histograms generated from the bootstrap distribution based on 5000 replications. The left subplots provide the frequencies out of the 5000 bootstrap replications for the percentage of correctly identified falsifiers, while the right subplots show the corresponding findings for the correctly assigned honest interviewers.

Starting with Figure 3.1 we see that the probability of correctly identifying all falsifiers is pretty low but slightly increases by decreasing the number of interviewers.

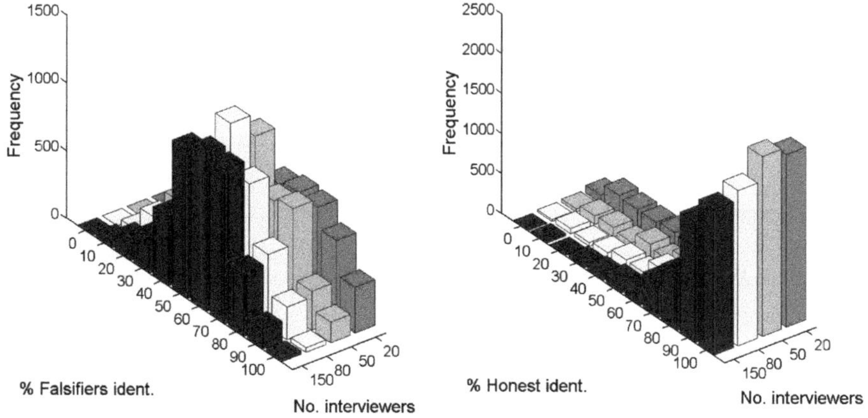

Fig. 3.1: Performance of clustering method as a function of number of interviewers for a substantial share of falsifiers (50%)

Figure 3.1 also shows that in most cases we are able to identify at least half of the falsifiers. This corresponds to the results obtained for the original dataset mentioned above and so we conclude that the results of the identification of the falsifiers still remain satisfying when decreasing the number of interviewers (as compared to the number of interviewers in the original dataset). This is also true for the identification of the honest interviewers. As in the original setup, we could identify nearly all honest interviewers and as the right part of Figure 3.1 shows, the probability of identifying all honest interviewers never falls short of 30%. Even for the smallest problem instance with 20 interviewers, all honest interviewers could be identified in about 45% of all bootstrap replications.

These findings remarkably improve with regard to the identified falsifiers in Figure 3.2 when the share of falsifiers is rather low (5%). The probability of identifying all falsifiers clearly increases. Including 150 interviewers we could identify all cheaters in about 25% of the bootstrap replications and when decreasing the number of interviewers the probability of correctly identifying all falsifiers increases up to 80% for the smallest problem instance including only 20 interviewers.

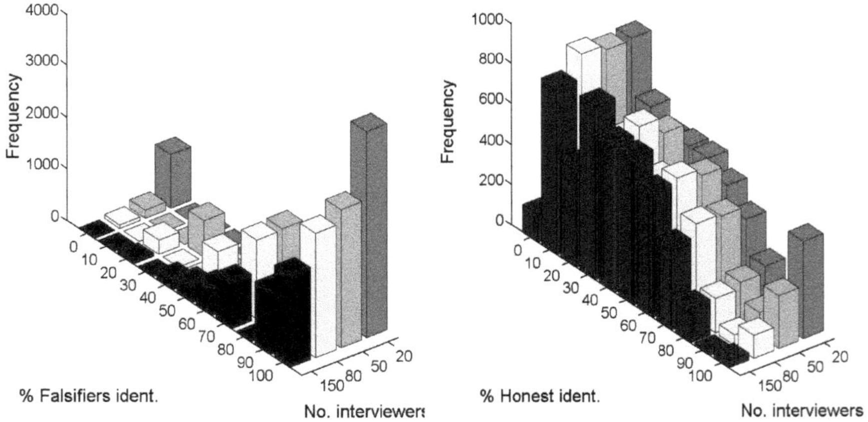

Fig. 3.2: Performance of clustering method as a function of number of interviewers for a small share of falsifiers (5%)

The right part of Figure 3.2 shows the result of the identification of the honest interviewers and reveals that the high share of correctly identified falsifiers comes at the cost of a high rate of false alarms. However, the rate of false alarms decreases with a decrease of the number of interviewers. Also the probability of identifying all honest interviewers increases but never reaches such high values as compared to Figure 3.1 when including a higher share of falsifiers. We can resume that the clustering method is robust against a decrease of the number of interviewers in the way that the results improve but it is sensitive to the share of falsifiers in the dataset. We investigate in more detail the effect of the share of falsifiers in a later section.

Number of Interviews per Interviewer

In our second experiment, we modify the number of interviews per interviewer. By means of the bootstrap method we resample 5, 20, 50, and 100 interviews out of all conducted interviews for every interviewer. The division of honest and falsifying interviewers is set to the same value as it occurs in our original data (50% falsifiers) and also the total number of interviewers remains unchanged (156 interviewers). Figure 3.3 shows the findings by means of boxplots where the left part of the figure displays the share of the correctly assigned falsifiers, while the boxplots in the right part show the corresponding results for the correctly identified honest interviewers.

It can be seen that the share of correctly assigned falsifiers varies between 40% and 70% with a median at about 55%, i.e. in more than half of all bootstrap replications 55% of all cheaters are correctly assigned. This finding is robust against a decrease of the number of interviews. Even in the case with 100 interviews per interviewer the median remains nearly unchanged. It might be concluded that the number of interviews has no strong effect on the frequency of correctly assigning the falsifiers in this experimental setting.

The right part of Figure 3.3 shows the result for the identification of the honest interviewers. We see that the results improve markedly. In the setting with only 5 interviews per interviewer we start with a median at 80%. With the increase of the number of interviews the median increases also up to 90% and at the same time the interquartile range becomes much smaller, i.e. the lower quartile increases from about 68%, for the setting with the lowest number of interviews to about 83%.

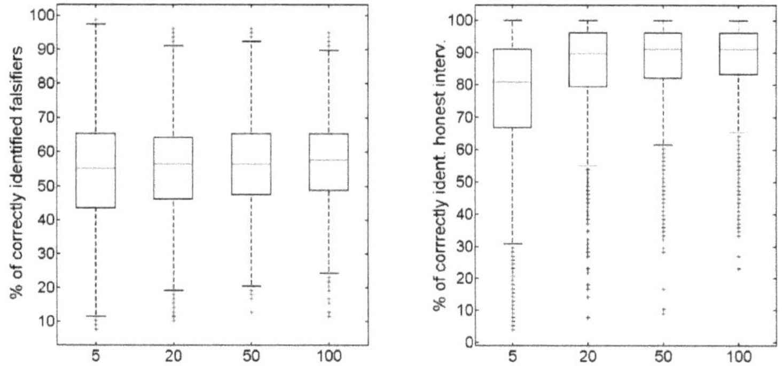

Fig. 3.3: *Performance of clustering method as function of number of interviews per interviewer (78 falsifiers)*

We can conclude that with regard to the performance of the clustering method, it would be favourable to have a high number of interviews per interviewer. The higher the number of interviews the higher is the share of correctly assigned honest interviewers. However, in practical settings, the number of interviews per interviewer will be limited for several reasons. Nevertheless, the results also indicate that a number of about 20 interviews per interviewer might be sufficient to obtain already a high discriminatory power for the questionnaire considered in this application.

Share of Falsifiers

In the first subsection, we already considered an experiment with different shares of falsified cases, but the focus there was on the influence of the total number of interviewers on the detection frequencies. In this step, we examine whether the clustering method still performs in a satisfying way when varying the share of falsifiers in the dataset, ranging from a very small (5%) to a very high value (95%). As Table 3.1 shows, the first setting corresponds to the original number of interviewers (N=156), which will be decreased to 80 and 20 in the remaining settings. As for the first experiments, the number of interviews per interviewer is always set to the same value as in the original dataset.

Figure 3.4 reports the results for the experiments with 156 interviewers. In the left plot, histograms for the percentage of correctly assigned falsifiers are shown for the different shares of falsifiers (9/147 (5%) to 148/8 (95%)). The right plot provides the corresponding results for the correctly identified honest interviewers. Overall, it can be summarized that the percentage of correctly assigned falsifiers is highest for very small shares of falsifiers. While this share of correctly assigned falsifiers decreases with an increasing share of falsifications, even for the highest share of falsifiers we could identify at least half of the cheaters in about 50% of the bootstrap replications.

This trend with regard to the share of correctly assigned falsifiers is reflected by the share of correctly assigned honest interviewers, which tends to increase with higher shares of falsifications. The right part of Figure 3.4 shows that only 10% of the bootstrap replications identify 100% of the honest interviewers if the share of falsifiers is just 5% (9 out of 156), while in 30% of the replications all 78 honest interviewers are correctly assigned if the dataset includes 78 falsifiers (50%). When including more than 50% of falsifiers the results deteriorate again.

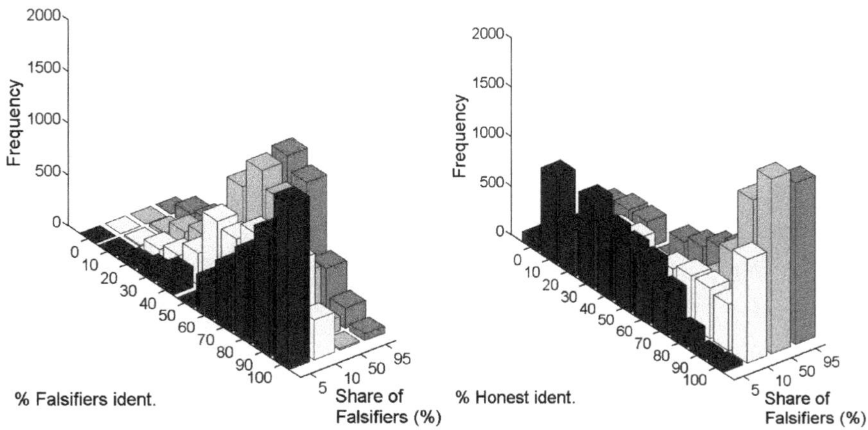

Fig. 3.4: *Performance of clustering method as a function of share of falsifiers (156 interviewers)*

For the other experiments with 80 and 20 interviewers we find that the results improve both for the identification of the cheaters and the honest interviewers. As the findings for the experiments with 80 interviewers are qualitatively similar, we do not report the figures to save space. Figure 3.5 shows the results for the setting with 20 interviewers.

The identification of the only falsifier (1/20, 5%) succeeds in this setting in about 75% of the bootstrap replications. This reflects the lower complexity of the task to cluster 20 interviewers correctly as compared to 80 or 156. By adding more falsifiers to the dataset (10% to 95%) with 20 interviewers, the clustering result deteriorates with regard to falsifications, whereas the findings for the identification of the honest interviewers improve. The right part of Figure 3.5 shows that for the largest problem instances with 95% falsifiers we assign the only honest interviewer correctly in nearly 70% of the replications.

Overall, it can be summarized that the falsifiers are most often correctly assigned if their share is low and share of correct assignments slightly increases with a lower total number of interviewers.

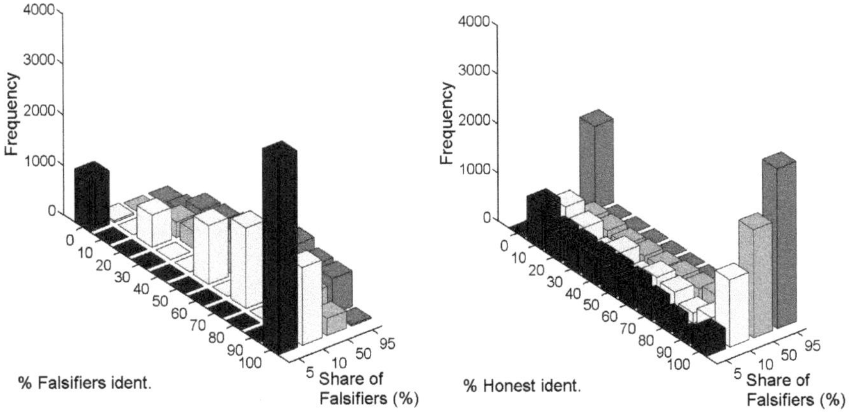

Fig. 3.5: *Performance of clustering method as a function of share of falsifiers (20 interviewers)*

Conclusion and Outlook

In this contribution we applied a bootstrap analysis to a method for the data driven identification of falsifications in survey data. The aim consisted in assessing the robustness of the method against specific modifications of the structure of the dataset. Due to the absence of sufficient available real data with ex-post known falsifications and different sample characteristics, we resort to create synthetic data based on an existing dataset from an experimental study.

The modification of the data characteristics is driven by the interest in the sensitivity of the method with regard to the total number of interviewers in the dataset, the number of interviews conducted per interviewer and the division of honest and cheating interviewers. The findings indicate that the promising results reported by Bredl et al. (2012) and also by Storfinger and Winker (2011) might not be considered as pure chance or statistical artefact, but seem to reflect an actual convincing performance of the proposed clustering method.

The applied bootstrap analysis shows that overall the clustering method is robust and that its performance even improves when decreasing the total number of interviewers. This is also true with regard to the increase of the number of interviews conducted by every interviewer whereby the share of correctly assigned interviewers even improves markedly. As mentioned before one of our main objectives is to examine the conditions for an increase of correctly identified cheaters. Our findings reveal that an increase is possible if the share of cheating interviewers and ideally the total number of interviewers is rather small. However, it has to be taken into account that the number of false positives, i.e., honest interviewers erroneously assigned to the cluster containing the falsifiers, might be substantial for some settings, e.g., very low shares of falsifiers and limited data per interviewer.

Future research in this field will have to address several issues. First, due to the fact that the cluster result depends on the choice of the indicators, this selection has to be investigated further. It may be the case that the used subsets of indicators might support only the increase of the share of correctly assigned honest interviewers whereby other subsets would produce a more discriminated cluster result, i.e. a high share of correctly assigned cheaters. Second, we still have to pursue the application of the identification method to further datasets. It could be favourable for the development of the method to test it with "real" falsifications which are not produced in an experimental setting but in real conditions as the data used in our previous work. Perhaps the share of correctly identified cheaters reaches as high values as in the previous analysis (Storfinger and Winker 2011). Third, until now we assumed that all falsifiers show the same behaviour when falsifying the interviews, e.g. the avoidance of the option "others" in semi-open questions. It might also be sensible to group the falsifiers into different groups depending on their motivation or strategy for cheating. Then, we have to take into account that the hypotheses concerning the direction of the indicators do not have to hold for all groups of cheaters, but rather for a specific type of falsifiers. Consequently, we have to consider more than two clusters produced by the cluster analysis.

Finally, in this contribution we dealt only with complete falsifications, in other words all falsifiers fabricated the whole questionnaire in all their interviews. Several studies (Hyman et al. 1954, Reuband 1990) showed that the more realistic way of cheating is to fabricate a part of the questionnaire. This means that the interviewers actually contact the respondent but conduct only a small part of the interview. The remaining part of the questionnaire will be done by the interviewers themselves using the information they already collected in this short interview. We suppose that the identification of partly falsified interviews by means of statistical tools is more difficult to manage than the identification of complete fabrications. Applying probabilistic clustering methods in this situation will be subject of future research. Thereby, every interviewer is characterised by a specific probability of being a cheater, and this information could be again used for further check-ups in the survey institutes. By means of the bootstrap method we will also be able to evaluate the performance of such a probabilistic clustering method when varying the share of fabricated data in each questionnaire.

Bibliography

Benford, Frank: The law of anomalous numbers. Proceedings of the American Philosophical Society 78(1), pp. 551–572, 1938.

Bredl, Sebastian, Peter Winker and Kerstin Kötschau: A Statistical Approach to Detect Interviewer Falsification of Survey Data. Survey Methodology 38/1, pp. 1-10, 2012.

Bredl, Sebastian, Storfinger Nina and Natalja Menold.: A Literature Review of Methods to Detect Fabricated Survey Data. In: Winker, Peter, Natalja Menold and Rolf Porst (Eds.): Interviewers' Deviations in Surveys – Impact, Reasons, Detection and Prevention. Peter Lang, Frankfurt a. M., 2013, pp. 3-24.

Chernick, Michael R.: Bootstrap Methods: A Guide for Practitioners and Researchers. Wiley, Hoboken, NJ. 2nd Ed., 2008.

Efron, Bradley: The Jackknife, the Bootstrap, and other Resampling Plans. Vol. 38 of CBMS-NSF Monographs, Society of Industrial and Applied Mathematics, 1982.

Hauck, Martin: Is survey postcard verification effective? Public Opinion Quarterly 33, pp. 117–120, 1969.

Hood, Catherine C. and John M. Bushery: Getting more bang from the reinterviewer buck: Identifying 'at risk' interviewers. In: Proceedings of the American Statistical Association (Survey Research Methods Section), pp. 820–824, 1997.

Hyman Herbert, William J, Cobb, Jacob J. Feldmann, Clyde W. Hart and Charles Herbert Stember: Interviewing in Social Research. Chicago, University of Chicago Press, 1954.

Jain, Anil K. and Jean-Vincent Moreau: Bootstrap technique in cluster analysis. Pattern Recognition 20(5), pp. 547–568, 1987.

Koch, Achim: Gefälschte Interviews: Ergebnisse der Interviewerkontrolle beim ALLBUS 1994. ZUMA-Nachrichten 36, pp. 89–105, 1995.

Menold, Natalja, Nina Storfinger, Peter Winker: Development of a method for ex-post identification of falsifications in survey data, Proceedings of New Techniques and Technologies for Statistics - NTTS 2011, Brussels, Belgium, 2011.

Menold, Natalja, Peter Winker, Nina Storfinger and Christoph J. Kemper: A Method for ex-post Identification of Falsifications in Survey Data. In: Winker, P., N. Menold and R. Porst (Eds.): Interviewers' Deviations in Surveys – Impact, Reasons, Detection and Prevention. Peter Lang, Frankfurt a. M., 2013, pp. 25-47.

Peck, Roger, Lloyd Fisher and John Van Ness: Approximate confidence intervals for the number of clusters. Journal of the American Statistical Association 84(405), pp. 184–191, 1989.

Reuband, Karl-Heinz: Interviews, die keine sind. "Erfolge" und "Mißerfolge" beim Fälschen von Interviews. Kölner Zeitschrift für Soziologie und Sozialpsychologie, 42(4), pp. 706-733, 1990.

Schäfer, Christin, Jörg-Peter Schräpler, Klaus-Robert Müller and Gert G. Wagner: Automatic identification of faked and fraudulent interviews in the German SOEP. Schmollers Jahrbuch 125, pp. 183–193, 2005.

Schnell, Rainer: Der Einfluss gefälschter Interviews auf Survey Ergebnisse. Zeitschrift für Soziologie 20(1), pp. 25–35, 1991.

Schreiner, Irwin, Karen Pennie and Jennifer Newbrough: Interviewer falsification in census bureau surveys. In: Proceedings of the American Statistical Association (Survey Research Methods Section), pp. 491–496, 1988.

Storfinger, Nina and Marie Kristin Opper: Datenbasierte Indikatoren für potentiell abweichendes Interviewerverhalten. Discussion Paper 58, ZEU, Giessen, 2011.

Storfinger, Nina and Peter Winker: Robustness of Clustering Methods for Identification of Potential Falsifications in Survey Data. Discussion Paper 57, ZEU, Giessen, 2011.

Ward, Joe H.: Hierarchical grouping to optimize an objective function. Journal of the American Statistical Association 58(301), 236–244, 1963.

Winker, Peter.: Optimization Heuristics in Econometrics: Applications of Threshold Accepting. Wiley, Chichester, 2001.

Detecting Poorly Conducted Interviews

Jörg Blasius and Victor Thiessen

Abstract

In their recent book, Blasius and Thiessen (2012) introduced several screening methods to assess the quality and validity of survey data. They characterized the survey interview context as one in which task simplification, time and effort minimization, and cost reduction strategies by respondents, interviewers, and research institutes resulted in poor data quality. In this paper, we concentrate on the quality of the interviewers, identifying patterns that help assess how carefully and thoroughly they conduct their interviews. We illustrate our ideas using the German General Social Survey 2008 in which we detect clusters of interviewer-specific response combinations whose frequency of occurrence defies the odds to such an extent that we suspect interviewer fraud to be the cause of some of them. Using two of the screening methods proposed by Blasius and Thiessen (2012), we find a substantial number of interviewers who simplified their tasks in a manner that reduced their interviewing time and effort but increased their "measurement error".

Introduction

In a recent study, Blasius and Thiessen (2012) proposed several methods to assess the quality of survey data. Using a variety of data sets, they documented that poor data quality accrued from three distinct sources: a) respondent task simplification strategies, b) interviewer techniques to minimize the length of time necessary to conduct each interview, and c) research institute practices to reduce their costs per interview. The research literature on response sets and other forms of respondent "satisficing" behavior (cf. Krosnick 1991, 1999, Krosnick and Alwin 1987) has deepened our understanding of the variety of ways in which respondents simplify the task of responding to survey questions in a manner that nevertheless fulfills their felt obligation to participate in the interview. Interviewers can reduce the time it takes to complete their interviews by faking, or partially faking, them. For example, they can skip some of the more time-consuming questions and fill in plausible responses later on. Research institutes can reduce their costs and increase their profit by the simple expedient of manufacturing (through copy and paste procedures) some interviews in order to obtain the contracted number of interviews. What the three sources have in common is that each of them represents a way of minimizing the time and effort necessary to complete interviews. And the consequence of all of these practices is to reduce the proportion of substantive variation in the data, which is our criterion for assessing its quality.

In this paper we focus only on interviewer tactics, discussing two techniques they might use to shorten the length of time it takes them to complete an interview. The first is to impute simple response patterns. For example, assume that there are ten questions on trust towards different institutions using ten point scales, running from "no trust at all" to "a lot of trust". In this case interviewers might ask about just one of the listed institutions and then generalize the response to all of them, perhaps by asking respondents whether they feel the same about other institutions generally. A simplified response pattern consists of ten times a "1" or ten times a "5" or a "10", or if there is one outlying institution such as banking to which respondents may have no trust at all, nine times a "10" and one time a "1". The second tactic is for interviewers to skip various parts of the interview during the face-to-face contact, especially long or complex sets of items, and to fill in the blanks at home. Also possible, but involving greater risk of getting caught, is to fake the entire interview.

We illustrate our ideas using the German General Social Survey 2008 (in German: ALLBUS2008) in which we detect clusters of interviewer-specific response combinations whose frequency of occurrence defies the odds to such an extent that we conclude that some of them occur because of interviewer fraud. Based on three different item sets contained in ALLBUS2008, we demonstrate how to detect anomalous patterns and how to determine that these unusual patterns were, in all likelihood, artifacts of the data collection procedure; in the given example mainly caused by some of the interviewers.

Theory

Since interviewers are typically paid on the basis of the number of interviews they have completed, it is in their monetary interest to cut corners to such an extent that constitutes fraud. To reduce possible interviewer fraud, research institutes must implement stringent quality control procedures. According to the American Association of Public Opinion Research (AAPOR; 2003:7), interviews should be monitored by the institute by re-contacting 5 to 15 percent of all respondents to ascertain whether the interview had indeed been conducted. This quality control procedure might be sufficient to detect completely faked face-to-face interviews. However, some interviewers, especially experienced ones, will be aware of such monitoring and will not risk faking entire interviews; they will fake only parts of them (Blasius and Friedrichs 2012, Blasius and Thiessen 2012). This can be done by actually contacting the respondents, asking them only a few of the simple questions, and then filling in the rest of the questionnaire at home by creating response patterns that the interviewer considers to be "plausible response combinations". From a rational-choice point of view, partially-faked interviews are perhaps optimal, since item batteries that take a relative long time to process as well as other time-consuming questions are fraudulently fabricated while basic and easy to verify data such as age and gender, are correctly collected. Such a strategy can shorten the interview time dramatically, which may please the respondent perhaps as much as the interviewer (cf. Blasius and Thiessen 2012).

Turning to the incidence of interviewer fraud, in a project of the U.S. Bureau of Census on detection of cheating behavior between September 1982 and August 1987 (cf. Schreiner, Pennie, and Newbrough 1988), Biemer and Stokes (1989: 25) reported that between three and five percent of all interviewers committed some form of cheating, whereof 72 percent were considered to be "complete fabrication of interviews". Stokes and Jones (1989: 696) pointed out that "it is generally believed that interviewers are most likely to curbstone in hard-to-enumerate areas, where the non-match rate is likely to be high." Bredl et al. (2012) reported a study in which all cases in one (out of two) site(s) were complete fabrications. Additionally, applying a multiple indicator based method, they were largely successful in discriminating between honest interviewers and those who fabricated their interviews. Schäfer et al. (2005: 10) reported up to 2.4 percent fakes for the German Socio-Economic Panel (GSOEP) for single subsamples in the first wave. Using a different approach, Nelson and Kiecker (1996) simply asked 173 telephone interviewers how fastidiously they follow their instructions. Even though they were supervised in their telephone laboratory, ten interviewers reported that they at least sometimes fabricate an interview and 22 said that they at least sometimes fabricate an item response (Nelson and Kiecker 1996: 1114). Harrison and Kraus (2002: 319) conclude "with the warning that interviewer cheating may be more prevalent than is generally assumed" and further "profit making institutions (which include universities to some extent) are certainly not interested in publications on interview fabrications, particularly regarding their own research" (Harrison and Kraus 2002: 321). As Blasius and Friedrichs (2012) note, fabricating parts of interviews in face-to-face surveys without being discovered by control mechanism such as those recommended by AAPOR (2003) is relatively simple.

Before demonstrating how to detect (partly) faked interviews and how to detect simplified response structures (see also Menold and Kemper and Winker et al., both in this volume), it may be helpful to raise several issues concerning fabricated interviews. The first is: When should an interview be classified as fabricated? Is an interview already faked when in random walk procedures, interviewers select the wrong household or fail to observe the selection rule within the household such as interviewing the person whose birthday was most recent?

According to AAPOR (2003: 2) "falsification includes: a) fabricating all or parts of an interview – the recording of data that are not provided by a designated survey respondent and reporting them as answers of that respondent; b) deliberately misreporting disposition codes and falsifying process data (e.g., the recording of a refusal case as ineligible for the sample; reporting a fictitious contact attempt); c) deliberately miscoding the answer to a question in order to avoid follow-up questions; d) deliberately interviewing a non-sampled person in order to reduce effort required to complete an interview; or e) otherwise, intentionally misrepresenting the data collection process to the survey management."

In a telephone laboratory, in principle one should be able to eliminate these types of interviewer fraud. However, in face-to-face interviews with random-walk instructions it is both time-consuming and costly to check whether rules such as *last birthday* and *every n^{th} household* have been observed. In the first case there is a high probability of obtaining a refusal when the interviewer initially asks for the birthdays of members of the household since the exact day of birth is a rather sensitive question, and it is well-known that one should avoid such questions at the beginning of an interview. Furthermore, when asking for the person *whose birthday is next*, the person who opens the door or who answers the phone call can easily name somebody who is currently not in, thereby providing a good excuse for not doing the interview. In the (unlikely) case the interviewer fulfilled this instruction correctly and the responded provided the correct information, the person selected by the last birthday criteria might have changed between interviewer and supervisor contact – and again, if supervisors are to fulfill this criteria, they would have to ask for the exact dates of birth of all household members. We are skeptical that many households would give this sensitive information to a stranger at the door or on the phone, especially in countries in which data security is a public concern. With respect to selecting the nth household, a third person would have to be sent to the sample points to confirm the random walk. Would research institutes routinely perform such time consuming procedures, and would they risk losing successful interviewers who provided well-done interviews just because they made a "mistake" when recording whose birthday was next or which was the nth household?

It can be stated without contention that at least in Western countries completely faked face-to-face interviews are relatively easy for institutes to detect using routine call backs; even the approximate age and gender can already be judged from the voice of the respondent. The detection of partly faked interviews is rather more complicated: the interviewers actually contacted the households and did ask some of the questions, and, barring stupidity, they made certain they asked those questions that are easy to verify, such as sex, age, and household composition. Assuming some rational behavior on the part of the interviewer, partly faked interviews would consist of fabricated responses to long item sets and other time consuming questions while basic and easy to verify data would be correctly recorded (Blasius and Friedrichs 2012, Blasius and Thiessen 2012, see also Bredl et al., this volume).

The second issue concerns the reasons and motives for faking (parts of) interviews. According to Crespi (1945: 431), the "cheater problem" is "essentially a moral one". He assumed that cheating "lies as much in the structure of the ballots and the conditions of administration as in the personal integrity of the interviewer" (Crespi 1945: 431). He discussed several problematic features of questionnaires that may demoralize interviewers; among others: unreasonable length, too many "whys" and "wherefores", apparent repetition of questions, lengthy wording, complex, difficult, and antagonizing questions. In addition to these reasons, Harrison and Krauss (2002: 325) hypothesize possible psychological rewards for faking such as "sense of cleverness" and "excitement of risky behavior". Although these may be good reasons and possible motives, it is also true that interviewers will inevitably increase their rate of pay per interview by opting to (partly) fake interviews rather than follow the instructions.

A third issue concerns the effects of faked interviews (see also Opper and Menold, this volume). Murphy et al. (2004) compared the response distributions each interviewer obtained on a set of questions about reported lifetime use of cigarettes, alcohol, marijuana, cocaine and heroin to identify interviewers whose distributions appeared to be highly unlikely, given their caseloads. The authors reported significant differences between falsifiers and non-falsifiers on some of them. In a subsequent paper, Murphy et al. (2005) extended the method to a more general data mining approach in which they searched for variable codes that any interviewers reported below or above the average of other interviewers. In a similar vein, Turner et al. (2002) found significant differences in the household composition and the number of lifetime partners between falsifiers and non-falsifiers. In a different tack, Porras and English (2004: 4223) used unlikely combinations such as "heavy smokers who also get considerable quantities of vigorous exercise" as a criterion to identify fraudulence on the part of the interviewer. The disadvantage of these methods is that the identification of falsifiers requires a large number of interviews from each interviewer. Further, "detections" because of "mistakes" such as using unlikely combinations can easily be avoided by applying our guide on "how to successfully fake an interview" (Blasius and Friedrichs 2012: 53). Further, an unlikely response combination is a rather risky indicator for a fabricated interview; it simply might be that the respondent was careless or failed to understand the question.

The final issue, and the one we address concretely, is how to detect faked interviews. One approach is that of Biemer and Stokes (1989: 23) who proposed "a general quality control sample design", which enabled them to estimate the probability that interviews of a certain interviewer were fabricated. Besides various assumptions such as the "probability that an interviewer is susceptible to cheating" (Biemer and Stokes 1989: 28) and other parameters about the interviewers and the sample, the model requires "a) a re-interview survey to detect interviewer cheating, b) an inspection of all interview forms by clerks who are specially trained to detect interviewer errors in completing the forms, and c) an annual on-site observation of the interviewer by a supervisory representative as the interviewer completes the assignment" (Biemer and Stokes 1989: 31). Their approach seems to be a sophisticated instrument for the detection of frauds, but it is both time consuming and probably not sufficient to detect faked sets of items.

Another possibility is the application of Benford's law, which is based on the empirical observation that many naturally-occurring tables of numerical data, for example, the distribution of income, exhibit leading digits that are not uniformly distributed as one might expect, following instead a certain logarithmic distribution (Benford 1938). Schäfer et al. (2005) applied this method to ascertain whether it could identify their known falsifiers in the German Socio-Economic Panel (GSOEP) data, and Bredl et al. (2012) applied a variant of this method to help identify data fabricators. A main disadvantage of this method is its limitation to numerical data such as income (recorded as whole numbers rather than grouped into categories); it cannot be applied to four- or five-point categories, which typically characterize survey data.

A third approach lends itself in particular to surveys that focus on specific groups (such as immigrants or visible minorities) or are conducted in neighborhoods with distinctive features. Blasius and Thiessen (2012) relied on the fact that interviewers have preconceived images about their respondents, especially when their respondents have distinctive characteristics or the interviews are collected in socio-economically homogeneous areas. In line with attribution theory, they hypothesized that the images interviewers attribute to their respondents would differ in systematic ways from the images that the respondents have of themselves. One such difference is that the interviewer's structured images contain stereotypic elements. This results in interviewer-respondent differences in the mean value of any attribute. It also results in less variability in the images that interviewers have of their respondents than the actual variability. That is, interviewers tend to see their "respondents" as more similar than they are. Schäfer et al. (2005) provide empirical support for this expectation (see also Porras and English 2004, Turner et al. 2002). A further stereotypic element is that interviewers are apt to see their respondents as being more consistent or predictable than they are. That is, the correlation between elements in their images is likely to exceed the correlation obtained from the genuine respondents.

Blasius and Thiessen (2012) capitalized on these hypothesized interviewer-respondent differences as part of their approach to screening data for its quality and evidence of fraud. In addition, they hypothesized that interviewers who faked their data would be marked by three further tendencies. First, since interviewers understand the meaning of the questions, they will not construct astonishing combinations of responses, which respondents sometimes do just because they misunderstood some questions. This too leads to greater consistency or correlation between related items. Second, Schäfer et al. (2005) and Bredl et al. (2012) showed evidence that fakers are more likely to provide substantive responses to every question, thereby minimizing the likelihood of missing data. Finally, interviewers who fake interviews are likely to choose what they consider to be "safe" responses, which typically means avoiding extreme response options. Taken together, these tendencies will result in faked interviews being systematically different from the genuine ones. Using a data set that included 35 fabricated interviews out of a total of 724, Blasius and Thiessen (2012: 69-74) successfully discriminated between genuine and fabricated interviews. For example, by applying categorical principal component analysis to different item sets they showed that the mean values of the factor scores from the fabricated interviews were significantly different from the genuine ones.

The advantage of our screening methods is that they do not require large samples from the individual interviewers – even a single interview might be sufficient to raise suspicion, and there is no need for any information concerning interviewer characteristics. The method can already be applied during field work on the basis of, say, 200 machine-readable interviews; suspicious interviews can be examined with respect to the sets of items that are identified as possibly fabricated, and falsified interviews (as well as falsifiers) can be excluded in a very early stage of field work.

While Blasius and Thiessen (2006, 2012) used their screening method primarily to compare the quality of responses between countries, here we will use this method to compare the quality of responses between interviewers. To this end, we concentrate on two features of the method. First, we identify and examine simple response patterns and assess how they are related to the interviewers. Second, we apply categorical principal component analysis to determine whether large deviations from the mean values of the latent scales are obtained from single interviewers. If these deviations in the scale values coincide with the use of simple response pattern, the quality of the respective interviews is judged to be very low. We demonstrate the methods on the basis of data from the ALLBUS2008, which contains an identifier for the interviewers.

Data

The ALLBUS (Allgemeine Bevölkerungsumfrage der Sozialwissenschaften, or in the international context, German General Social Survey), is one of the most prominent social surveys in Germany. ALLBUS collects data on attitudes, behavior and social structure. Since 1980, every two years a representative cross section of the population is interviewed with a partly constant and partly variable question program (http://www.gesis.org/allbus, accessed on September 23, 2012). In 2008, the number of respondents was 3,469, with a response rate of 40.3%, from a total of 162 sample points in 148 communities, with an oversampling in the former East Germany (Wasmer, Scholz and Blohm 2010). The data were collected via face-to-face standardized interviews (CAPI – Computer Assisted Personal Interviews). In theory, the deployment of laptops should prevent interviewers skipping parts of the interviews to fill them in later since the input time of each response can be recorded easily. However, if either the interviewers are unaware of this possibility (or they are willing to risk detection) or the institute fails to check the time stamps, partly fabricated interviews might occur in the data set.

To search for (partly) fabricated interviews we apply our screening method to three different sets of items. The first set contains 12 political attitude items, with five response categories ranging from "strongly agree" to "strongly disagree" (see Table 4.1 for question wording). The second item set gauges trust in 13 institutions: Public health, Federal Constitutional Court, Bundestag (Parliament), City and municipal administration, Judiciary, Television, Press, Colleges and Universities, Federal Government, Police, Political parties, European Commission, European Parliament. Seven-point response options were provided, with the end points consisting of "no trust at all" and "a lot of trust". The third set contains nine four point items on attitudes towards politics and politicians, which is a variation of the well-known political efficacy and trust scale (Balch 1974). The categories range from "strongly agree" to "strongly disagree". The wordings of the questions are given in Table 4.4. In all three data sets, the non-substantive response option "don't know" was available.

The Detection of Simple and Outlying Response Patterns

It should not require detailed discussion to conclude that the use of simple response patterns such as repeatedly giving the same response to a set of items (for example, "strongly agree") constitutes a form of simplification, either by the interviewer or by the respondent. Using as an example the 13 trust in institutions items, interviewers can simplify their task by asking one or two general questions about respondents' trust in institutions or to one or two specific institutions and then assign that response to all (or large majority) of them. Likewise, respondents could simplify their task by stating that they (dis)trust all institutions to an equal extent after responding to just a few of the institutions; in Krosnick`s (1991, 1999) terminology, this is a possible form of satisficing behavior. Finally, there could be an implicit agreement between interviewers and respondents to shorten the set of questions towards institutions. For the questions regarding political attitudes as well as political efficacy and trust, an outlying response pattern would be, for example, total agreement with the positively-formulated items and total disagreement with the negatively-formulated ones. In the event the interviewer simplifies his/her task, there should be an association between interviewer and the likelihood of use of simple and outlying response patterns, i.e., a few interviewers should be responsible for a disproportionate majority of simple and/or outlying response patterns. Regardless of whether the respondent or the interviewer simplifies the task, the use of simple pattern of responses is an indicator of poor data quality.

As Blasius and Thiessen (2012: 61) show for the most recent World Value Survey (2004-2008), the use of simple response patterns is highly correlated with country. Among other anomalies, the authors could show for Germany that in a set of ten 10-point variables towards essentials of democracy (for which there are 10^{10} possible response patterns), a single response pattern occurred 128 times out of 1,839 interviews (= 7.0% of the sample, after excluding missing data in the given set of items). This is a clear indicator that at least some of the interviewers performed their job in a decidedly substandard manner.

The first item set we use consists of 12 variables, each with five response categories, which permit a total of 5^{12} = 244,140,625 possible response combinations. We will use this set of items to search for a) simple response patterns and b) outlying response structures. In recognition of the ordinal nature of the data, we apply categorical principal component analysis (CatPCA) rather than traditional principal component analysis (PCA) for the second task. CatPCA has the advantage that the distances between successive categories will be recalculated within an iterative procedure (Gifi 1990, De Leeuw 2006). Since this re-calculation can result in tied categories (i.e., two, three or even more successive categories are rescaled to have equal values), we cannot use this procedure to perform the first task; for this task we use PCA. Table 4.1 shows the wording of the questions for the first set of items as well as the factor loadings produced by CatPCA.

Table 4.1: Political opinions on Germany, factor loadings

Questions	F1	F2
Immigrants should be compelled to accommodate to the German customs and traditions	0.525	-0.484
Politicians should stay out of the economy	0.063	-0.346
Harsher measures should be taken to ensure environmental protection	0.429	0.338
Same-sex marriages should be legally recognized	0.156	0.616
Applications and promotions of women and men should undergo equal treatment	0.408	0.446
Offenders should be punished more severely	0.586	-0.371
Social security should be the prime objective of government policy	0.708	-0.001
Income and wealth should be redistributed in favor of the common people	0.668	0.007
Immigrants are good for the economy	-0.219	0.548
Germany should provide military support for the war against terrorism	-0.085	0.055
The individual citizens' rights and freedoms should be observed in all circumstances	0.435	0.335
The further opening of the world markets generates wealth for everyone	0.317	0.138

An inspection of Table 4.1 suggests that the first dimension has little substantively-interpretable meaning, and is best considered to be a form of response behavior (10 out of the 12 items are positively correlated with the first dimension, the remaining two weakly negative); the second dimension reflects some substantive meaning (conservative/traditional vs. liberal/modern). Although varimax rotation provides a different picture (the angle of rotation is 54°/36°), it also is not amenable to a substantive interpretation. Indeed, after rotation the first dimension suggests a response set interpretation even more strongly, since now only the item "politicians should stay out of the economy" is slightly negatively correlated with the first dimension.

Although the variables are inter-correlated and although several million combinations might be very unlikely, still on purely probabilistic grounds, the presence of multiple respondents with identical response patterns should be rather rare. As a quick procedure for screening these items, we applied PCA to the 12 variables using listwise deletion. We saved the factor scores for the first and second dimension, keeping in mind that in PCA identical response combinations provide identical factor scores. Note that missing responses could actually be included easily as valid responses; if this was done, the measurement level of the single items would obviously be even farther away from fulfilling the metric assumption, with the consequence that the resulting factors scores would not be meaningful at all. However, since we would have been interested only in isolating identical response patterns and not in any substantive solution we could have done it. We first plot the obtained frequency distribution of the factor scores as bar chart (Figure 4.1).

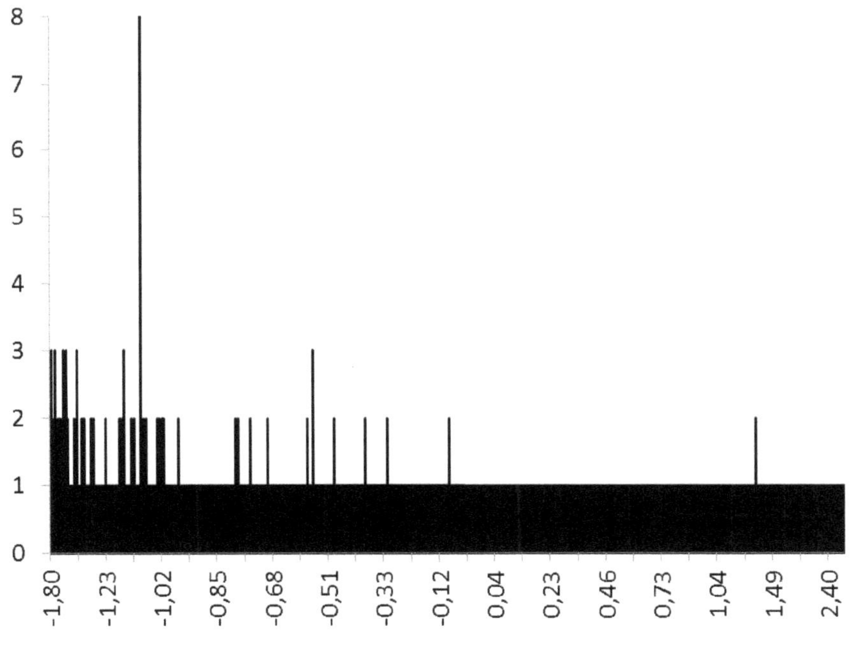

Figure 4.1: Bar chart of responses, first dimension of factor scores.

In Figure 4.1, the x-axis represents the factor scores while the y-axis gives the frequencies of the distribution. As previously mentioned, since more than 244 million response patterns are possible, factor scores that occur more than once are anomalous from a purely probabilistic point of view. With respect to the frequency distribution of equal responses, Figure 4.1 shows that one combination was used eight times. In addition, seven combinations appeared three times and 41 combinations occurred twice. The equal response patterns are concentrated on the left side of the scale, starting with two times responses of 1-1-1-3-1-1-1-1-5-5-1-1, followed by three times of 1-1-1-5-1-1-1-1-5-5-1-1, followed by two different response sets, followed by two times of 1-1-1-1-1-1-1-1-4-4-1-

1. The response structure that appeared eight times (1-1-1-1-1-1-1-1-1-1-1-1) is the simplest response structure possible and is devoid of any substantive meaning (see the questions displayed in Table 4.1). In the next section we will assess whether the last response structure occurs disproportionately often in interviews conducted by a given interviewer.

The second example consists of 13 seven-point-variables towards trust in institutions. The univariate descriptions of these items do not reveal anything particularly astonishing: political parties appear to be trusted the least while the police and the Federal Constitutional Court are awarded the greatest trust; the highest percentages of missing values belong to the colleges and universities as well as to the European Commission and the European Parliament (see Table 4.2).

Table 4.2: Trust in institutions

Institution	Mean value	Median	Standard Deviation	% Non-Response
Public health	4.02	4	1.52	0.6
Federal Constitutional Court	4.78	5	1.52	9.5
Bundestag	3.64	4	1.41	3.9
City and municipal administration	4.40	4	1.47	1.9
Judiciary	4.21	4	1.51	2.7
Television	3.56	4	1.43	1.3
Press	3.93	4	1.37	2.6
Colleges and universities	4.91	5	1.24	12.3
Federal Government	3.67	4	1.42	1.7
Police	4.80	5	1.37	1.0
Political parties	3.10	3	1.32	3.1
European Commission	3.36	3	1.43	13.1
European Parliament	3.38	3	1.47	12.6

1 = no trust at all, 7 = very great trust

Table 4.3: Trust in institutions, frequencies of the statements (in percent)

A lot of trust			Neither … nor			No trust at all	Category
1,952	5,254	9,015	10,804	8,298	4,570	2,928	N
4.6	12.3	21.1	25.2	19.4	10.7	6.8	N (in %)
73.3	37.2	15.6	10.6	20.5	47.7	71.8	N=0
14.6	23.0	19.1	14.7	19.1	17.7	10.2	N=1
6.0	17.0	19.3	18.5	18.4	14.2	5.8	N=2
2.5	10.4	16.7	17.5	15.8	9.2	3.9	N=3
1.6	5.7	12.6	15.1	11.1	4.9	2.7	N=4
0.5	3.2	8.0	10.3	6.6	3.3	1.6	N=5
0.4	1.8	4.4	5.9	4.3	1.4	1.4	N=6
0.5	0.9	1.9	3.6	2.6	0.9	0.7	N=7
0.2	0.3	1.3	1.7	0.7	0.3	0.7	N=8
0.1	0.2	0.6	1.0	0.5	0.2	0.5	N=9
0.1	0.1	0.2	0.5	0.2	0.1	0.3	N=10
0.1	0.1	0.1	0.3	0.1	0.1	0.1	N=11
0.1	0.1	0.2	0.1	0.0	0.0	0.1	N=12
0.2	0.0	0.1	0.2	0.0	0.1	0.2	N=13

In the next step we count the number of responses of each category over the 13 institutions (Table 4.3). In total, 42,821 responses were given by the 3,469 interviewees towards trust in institutions; on average, each respondent gave 12.3 valid answers. More than a quarter of all responses occurred precisely at the middle or neutral point of the scale, while the two extreme values "no trust at all" and "a lot of trust" were chosen relatively infrequently (6.8% and 4.6%, respectively). Notably, 71.8% of the respondents gave not a single "no trust at all", i.e., there is no single institution to which these respondents have no trust at all. Further, 10.2% of the respondents totally distrusted exactly one institution, and an additional 5.8% totally distrusted two institutions. At the other extreme, four respondents (= 0.1%) totally distrusted 11 institutions, another four (= 0.1%) totally distrusted 12 institutions, while six respondents (= 0.2%) provided a "no trust at all" to all 13 institutions. With respect to the mid-point of the scale ("neither-nor"), 10.6% of the respondents never used this response category, 14.7% used it just once, while eight respondents (= 0.2%) assigned this category to all institutions. Further, 73.3% of the respondents did not give any "a lot of trust" and eight respondents (= 0.2%) reported to have a lot of trust in all 13 institutions. In our opinion, providing identical responses towards all institutions represents a form of task simplification, either on the part of the interviewers or the interviewees. In the given case, there are 27 interviews in which all institutions were judged identically, and there are 80 interviews in which at least 11 institutions were judged equally. The 27 interviews with identical responses towards all of the listed institutions will be the main criteria for selecting those interviewers who might have simplified their tasks (see next section).

For the final example we use the nine four-point items on attitudes towards politics and politicians; the wording of the questions and the CatPCA factor loadings for the first two dimensions are given in Table 4.4. As this table shows, the first dimension contrasts the positively-formulated items from the negative ones, while the second one contrasts items 2, 3, and 7 from the others.

Table 4.4: CatPCA factor loadings for attitudes towards politics

Question	F1	F2
All in all, life is very good in a country like Germany	-0.327	0.608
Politicians are hardly interested in the opinions of people like me	0.474	-0.455
I believe I am able to take an active role within a group that addresses political issues	-0.625	-0.266
The whole political complex is too complicated, so someone like me cannot understand what is actually going on	0.769	0.243
Politicians usually persevere to defend the interests of the people	-0.415	0.629
In general, I do not know much about politics	0.736	0.371
Most people would be fully able to participate and work in a political group	-0.268	-0.217
Average citizens like me only understand little about politics	0.465	0.295
In a democracy, it is the duty of every citizen to regularly participate in the elections	-0.323	0.454

In addition to the CatPCA factor loadings, we are also interested in the frequencies of the various response patterns. To detect them we performed a PCA using listwise deletion; the frequencies of the distribution of the factor scores (first dimension) are displayed in Figure 4.2.

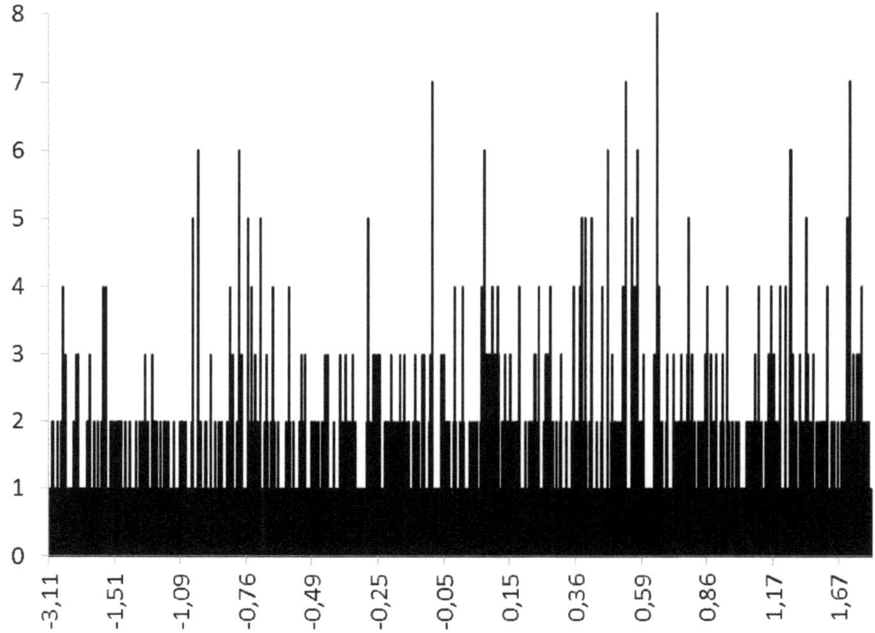

Figure 4.2: Distribution of factor scores (first dimension) of attitudes towards politics

Considering that more than 260,000 different response patterns are possible, Figure 4.2 shows that there are many instances of identical combinations. More specifically, 2,120 response patterns appear exactly once, 251 appear twice, 74 appear three times, 32 appear four times, 11 appear five times, seven appear six time, three appear seven times and one response pattern appears eight times. In other words, in 3,098 interviews without missing responses towards the political efficacy and trust items, we found 2,499 different response patterns, or, on average, each pattern appears 1.24 times; the item set comprises almost 20% of duplicates. Comparing this solution with solutions from other data sets (cf. Blasius and Thiessen 2012), any substantive interpretation based on this item set should be exercised with caution.

Interviewers, Respondents, and Task Simplification

In the final step, we associate the shown response patterns with the interviewers, using the identification number that is provided in the data set. Additionally, we compute the mean values of all interviewers on the latent scales from the "political attitudes" and the "political efficacy and trust" item sets. As our main criteria for selecting the subset of interviewers who might have simplified their task, we use the item set on trust in institutions. Here we found 27 "respondents" judging all institutions equally. In addition, we show the solutions for one interviewer with especially conspicuous values with respect to the distribution of the latent scales on both political attitudes and political efficacy and trust. The findings for the interviewers are shown in Table 4.5.

Table 4.5: Interviewer and response sample

Interviewer	Number of interviews	PA-resp. pattern (N=8)	PA-F1	PA-F2	Inst.-13 identical responses (N=27)	Inst. - min. 11 id. resp. (N=80)	Pefficacy-F1	Pefficacy-F2
122424	16		0.896	-0.026	1	1	0.662	-0.203
198749	17		0.090	0.153	1	1	0.431	-0.231
213616	56	1	-0.157	-0.042	1	2	-0.090	-0.598
236608	32		-0.187	-0.261	4	5	0.037	0.543
246883	19		-0.133	0.549	1	2	-0.425	-0.562
250180	30		-0.803	1.210	0	0	-0.898	1.108
268232	66		-0.044	-0.074	2	3	-0.263	0.089
271632	134		0.558	-0.151	1	2	0.282	-0.202
280283	30		-0.025	0.370	1	1	-0.223	0.205
280720	22		0.479	-0.187	1	2	-0.289	-0.341
281360	46	3	-0.730	-0.335	7	12	-0.927	-0.746
285382	11		-0.913	0.623	2	2	-0.551	1.515
291277	49		-0.088	-0.099	1	3	-0.214	0.099
298204	44		-0.523	0.420	1	2	-0.191	0.131
304280	32		0.507	0.099	1	1	-0.692	-0.047
304425	60	1	-0.294	-0.123	1	1	-0.127	-0.142
305553	6		1.219	1.054	1	1	-0.301	0.696

Table 4.5 shows the solutions from 17 interviewers who completed in total 670 of all interviews (= 19.3%) of all interviews. The first column shows the interviewer identification number, the second the numbers of interviews he/she conducted, the third the number of identical response patterns concerning the 12 questions towards political attitudes (pattern: 1-1-1-1-1-1-1-1-1-1-1-1, which appeared eight times in total). The fourth and fifth columns show the mean values of the interviewers on the first two dimensions on political attitudes, the sixth the number of identical responses towards the 13 institutions (our main criteria for the selection of interviewers, of which there were 27 instances), the seventh column shows similar patterns of responses towards the 13 institutions (80 cases in which at least 11 institutions were judged identically) and the last two columns provide the mean values of the interviewers on the nine items towards political efficacy and trust.

We start with our main criteria and assign the 27 interviews with unique responses towards the 13 institutions to the interviewers. Table 4.5 shows that interviewer "281360" has seven cases (out of 46 interviews), interviewer "236608" has four out of 15 cases, interviewers "268232" and "2285382" have two cases (out of 66 and 11, respectively), and another 12 interviewers obtained exactly one of those unique response patterns. While a single interview with identical responses towards all 13 institutions might equally be attributable to task simplification behavior by either the respondent or the interviewer (for example, interviewer "271634" with one out of 134 interviews), obtaining four or seven such response patterns is a clear indicator that it was the interviewer who simplified the task. Further, the likelihood for interviewer "285382", who completed only 11 interviews, to have two such unique response patterns is very low, while two out of 66 interviews (interviewer "268232") might just be coincidence. To obtain more evidence to strengthen the argument that particular interviewers simplified their task, we show that the same interviewers also provide conspicuously outlying values on other item sets.

Taking the 80 cases where we used a more relaxed criterion that permitted a maximum of two institutions to be judged differently provides the following distribution: interviewer "281360" has 12 out of 46 interviews, "236608" has five out of 15 cases, "296732" has four cases (from 63), "268232" and "291277" have three cases (from 66 and 49 interviews, respectively), and another ten interviewers have two cases. To summarize, three out of the four interviewers who obtained more than one unique response pattern using the stringent criterion, have even more similar response patterns when using the more relaxed criterion. With respect to the eight instances of the most simplified response patterns on the 12 items on political attitudes that are substantively meaningless, five of them occur among the interviewers already identified with the previous criteria. The "winner" with three such response patterns is again interviewer "281360".

In the remaining columns we show the mean values for the individual interviewers on the four latent scales described above. It should be reiterated that the mean values of these scales are standardized to a mean of zero with standard deviations of one. If we consider that we have face-to-face interviews and that the response structures are likely to be associated with the sample points via educational level and other socio-demographic characteristics, mean values between -0.3 and +0.3 might be regarded without any suspicions – and, considering only those interviewers who conducted more than 10 interviews, a great majority of all interviewers provided values within that range. To be on the safe side, in the following we comment only on mean values that exceed 0.50 in absolute value.

Starting with interviewer "122424", in addition to the one unique response towards the 13 institutions, he/she has outlying values on the first dimensions of both scales we considered for these analyses, i.e., he/she produced a large number of response sets towards political attitudes (PA-F1), and with respect to the items on political efficacy and trust (Pefficacy-F1) he/she is very positive. In this case we can conclude that this interviewer simplified his/her task, perhaps not in all but in a substantial number of cases. In some contrast, with a similar number of completed interviews (17 instead of 16) and also one unique response pattern, the mean values of interviewer "198749" on the latent scales do not raise much suspicion; the relatively high value on the first dimension of the political efficacy and trust scale (0.431) might be explained with the sample point – but this should have been checked by the institute after receiving the interviews. Without contrasting this value with the socio-demographic characteristics of the respective respondents, we could explain the equal responses towards the institutions with respondent tasks simplification behavior (to be sure, interviewer training is required to avoid such simplified responses).

With respect to the mean values of the latent variables, interviewer "250180" (30 interviews) provides the most outlying values; he/she must have a very distinct idea of how respondents might answer the corresponding questions. Although, he/she did not provide unique responses towards the 13 institutions, his/her "solution" on this set of items is markedly different from the average. From his/her 390 judgments towards the institutions (13 × 30 respondents), he/she did not have any missing values (on average, each respondent has 0.66 missing values; i.e. 20 missing values would have been expected), 19.7% of the responses are "no trust at all" (compared to 6.8% of the entire sample, cf. Table 4.3). In 20.0% of all cases his/her "respondents" gave the second lowest trust (average is 10.7%), and on the other side of the scale his/her "respondents" gave in 5.9% of the cases the second highest amount of trust and in 2.8% of the cases "a lot of trust", compared to 12.3% and 4.6% for the entire sample. In short, we are short of trust in this interviewer and suggest excluding his/her interviews from the data set.

There are only two interviewers whom we would trust even less, namely "281360" and "285382". The first of these two has a large number of simple responses to both sets of questions we considered for these analyses, and he/she has suspicious outlying mean values on three of the four latent scales. With only 11 interviews the other interviewer managed to obtain two unique response patterns towards the institutions and had extreme outlying responses towards the other two sets of items; a mean value of 1.52 (Pefficacy-F2) is especially "astonishing".

Another "interesting case" is interviewer "305553" who conducted only six interviews but obtained mean values exceeding 1.0 on the first two dimensions on political attitudes, and he/she has one unique response pattern towards the 13 institutions. In this case there is a possibility that some of the respondents were aware of each other (they are probably from the same sampling point) and agreed to give "funny answers" to confound the interviewer/the survey. Perhaps in this case the interviewer should have terminated the interview because of the astonishing answers. There are other interviewers we could comment on, but we leave this up to the reader. However, it should be noted that there are other interviewers not listed in Table 4.5, many of them having conducted only a small number of interviews, who obtained suspicious values on the latent scales. To summarize, most of the interviewers listed in Table 4.5 simplified their tasks considerably, especially interviewers "281360", "285382", and "250180".

Conclusion

Blasius and Thiessen (2012) extended the satisficing theory of Krosnick (cf. Krosnick and Alwin 1987, Krosnick 1991, 1999) by adding interviewers as well as research institutes into their theoretical model; they summarize it under the term "task simplification". While on the level of the respondents "satisficing" and "task simplification" are hard to distinguish, interviewers can only simplify their task either by fabricating the interviews or by (drastically) shortening the set of items, which often results in simple and/or outlying response patterns. If respondents simplify their task, there should be no association between simple and outlying response patterns and interviewer. On the other hand, if interviewers simplify their task, this should be manifested in a disproportionate number of simple and outlying response patterns in the interviews conducted by them. As shown in the paper, quite a few interviewers simplified their tasks involving a substantial number of interviews.

In identifying interviewers who took unwarranted short cuts, it does not mean that all of their interviews are simplified – but it is a number that likely affects the solutions in multivariate data analysis. However, it is not possible to prove in every instance whether it is the interviewer or the respondent who produced the simplification. If the interviewer is responsible, institutes need to monitor for simple and outlying response patterns more thoroughly and earlier. Indeed, this should be done relatively early during the field work, since 200 interviews should be sufficient to employ the proposed screening techniques as a way to detect suspicious interviews/interviewers. Finally, a discussion needs to be started on possible criteria for excluding interviews exhibiting exceptionally low quality, for example, when they provide response patterns that fail to exhibit any substantive meaning.

Bibliography

AAPOR (American Association for Public Opinion Research): Interviewer Falsification in Survey Research: Current Best Methods for Prevention, Detection and Repair of Its Effects (www.aapor.org/pdfs/falsification.pdf), 2003.

Balch, George I: Multiple Indicators in Survey Research: The concept 'sense of political efficacy. Political Methodology 1, 1-43, 1974.

Benford, Frank: The Law of Anomalous Numbers. Proceedings of the American Philosophical Society 78, 551-572, 1938.

Biemer, Paul P. and S. Lynne Stokes: The Optimal Design of Quality Control Samples to Detect Interviewer Cheating. Journal of Official Statistics 5, 23-39, 1989.

Blasius, Jörg and Jürgen Friedrichs: Faked Interviews, In: Samuel Salzborn, Eldad Davidov und Jost Reinecke (eds): Methods, Theories, and Empirical Applications in the Social Sciences, Festschrift für Peter Schmidt (49-56), Wiesbaden: Springer VS, 2012.

Blasius, Jörg and Victor Thiessen: Assessing Data Quality and Construct Comparability in Cross-national Surveys. European Sociological Review 22, 229-242, 2006.

Blasius, Jörg and Victor Thiessen: Assessing the Quality of Survey Data, London: Sage, 2012.

Bredl, Sebastian, Peter Winker and Kerstin Kötschau: A Statistical Approach to Detect Interviewer Falsification of Survey Data. Survey Methodology 38/1, 1-10, 2012.

Crespi, Leo P.: The Cheater Problem in Polling. Public Opinion Quarterly 9, 431-445, 1945.

De Leeuw, Jan: Nonlinear Principal Component Analysis and Related Techniques, In: Michael Greenacre and Jörg Blasius (eds.): Multiple Correspondence Analysis and Related Techniques (107-133), Boca Raton, Fl.: Chapman & Hall/CRC, 2006.

Gifi, Albert: Nonlinear Multivariate Analysis, New York: Wiley, 1990.

Harrison, David E. and Stefanie L. Krauss: Interviewer Cheating: Implications for Research on Entrepreneurship in Africa. Journal of Developmental Entrepreneurship 7, 319-330, 2002.

Krosnick, Jon A.: Response Strategies for Coping with the Cognitive Demands of Attitude Measures in Surveys. Applied Cognitive Psychology 5, 213-236, 1991.

Krosnick, Jon A.: Survey Research. Annual Review of Psychology 50, 337-367, 1999.

Krosnick, Jon A., and Duane F. Alwin: An Evaluation of a Cognitive Theory of Response-order Effects in Survey Measurement. Public Opinion Quarterly 51, 201-219, 1987.

Murphy, Joe, Rodney Baxter, Joe Eyerman, David Cunningham and Joel Kennet: A System for Detecting Interviewer Falsification, Paper presented at the "American Association for Public Opinion Research, 59th Annual Conference", May 13-16, Phoenix, Arizona: 4968-4975, 2004.
(www.amstat.org/Sections/Srms/Proceedings/y2004/Files/Jsm2004-000517.pdf).

Nelson, James E. and Pamela L. Kiecker: Marketing Research Interviewers and Their Perceived Necessity of Moral Compromise. Journal of Business Ethics 15, 1107-1117, 1996.

Porras, Javier, and Ned English: Data-driven Approaches to Identifying Interviewer Data Falsification: The case of health surveys, Paper presented at the "American Association for Public Opinion Research, 59th Annual Conference", May 13-16, Phoenix, Arizona, 2004.
(http://www.amstat.org/sections/SRMS/proceedings/y2004/files/Jsm2004-000879.pdf).

Schäfer, Christian, Jörg-Peter Schräpler, Klaus-Robert Müller, and Gert G. Wagner: Automatic Identification of Faked and Fraudulent Interviews in the German SOEP. Schmollers Jahrbuch 125, 183-193, 2005.

Schreiner, Irwin, Karen Pennie and Jennifer Newbrough: Interviewer Falsification in Census Bureau Surveys, Proceedings of the Survey Research Method Section, ASA, 1988: 491-496 (www.amstat.org/sections/srms/proceedings), 1988.

Stokes, Lynne and Patty Jones (1989): Evaluation of the Interviewer Quality Control Procedure for the Post-Enumeration Survey, Proceedings of the Survey Research Method Section, ASA, 1989: 696-698.
(http://www.amstat.org/sections/SRMS/Proceedings/papers/1989_127.pdf.)

Turner, Charles F, James N. Gribble, Alia A. Al-Tayyib, and James R. Chromny: Falsification in Epidemiologic Surveys: Detection and Remediation, Technical Papers on Health and Behavior Measurement (Report #53), 2002.
(http://dragon.soc.qc.cuny.edu/Staff/turner/TechPDFs/53_Falsify.pdf).

Wasmer, Martina, Evi Scholz and Michael Blohm: Konzeption und Durchführung der „Allgemeinen Bevölkerungsumfrage der Sozialwissenschaften" (ALLBUS) 2008. Technical Reports 2010/04, 2010.
(www.gesis.org/fileadmin/upload/forschung/publikationen/gesis_reihen/gesis_methoden berichte/2010/TechnicalReport_10-4.pdf).

II Detection and Prevention of Falsification in Survey Practice

Concepts and Practices in Interviewer Qualification and Monitoring

Birgit Jesske

Abstract

The main goal of quantitative survey research is collecting data under standardized conditions. The use of interviewers in particular is known to bring with it the risk of errors, variable behavior and uncontrollable interview situations. Standards and rules for interviewing are intended to guide the interviewers' behavior within the interview situation. These can be imparted by means of intensive training sessions and extensive training materials. Interviewer compliance can be ensured by means of a monitoring process throughout the data collection phase. This monitoring work allows the institute to quickly identify and remedy any deviant behavior or errors that may occur. This article introduces the standards and standard methods used for interviewer training and monitoring at infas – the Institute for Applied Social Sciences. They are designed to flexibly meet the requirements of complex surveys for scientific purposes. They have their origins in the experience and standards for surveys conducted by infas and are being continually adapted and improved.

Background – Standardised Data Collection

The primary aim of a standardised approach to collecting survey data is to ensure the validity and reliability of the results. The standardisation therefore needs to cover not only the individual wording of questions and the structure of the survey instruments but also their implementation and the way the interviewers address the respondents. The interviewing standards[1] stipulate how to present a question to the respondent, which explanations interviewers are allowed to give to the respondent, and what to do if the respondent refuses to answer. The interviewers' compliance with the standards is a key prerequisite for avoiding interviewer effects and errors, which in turn ensures the validity and reliability of the collected data.

The present article demonstrates how requirements for standardised interviewer behaviour can be imparted to the interviewers in practice and the ways in which compliance with the standards can be monitored. Standardising interviewer behaviour mainly concerns the processes of *contacting* the target person, i.e. processing all the addresses in the gross sample, and *conducting the interview*. A mandatory set of rules and corresponding training techniques define these tasks for the interviewers. Compliance with these rules is guaranteed by mon-

1 Various reference publications deal with interviewing standards – amongst others Prüfer/ Stiegler (2002), Fowler/ Mangione (1990).

itoring and control processes. Continuous, well-established routines are designed to ensure immediate detection and correction of irregularities.

A broad range of different concepts and strategies is available to guarantee standardised data collection and interviewer behaviour for surveys. This article will focus on four aspects:

- Interviewer *qualification* via extensive training techniques.
- Interviewer *motivation* through stimuli (also including payment).
- Continuous *monitoring* of interviewers' behaviour with regard to contacting and interviewing.
- Comprehensive *reporting* allowing the immediate detection and correction of deviant behaviour or errors.

These aspects and their implementation by infas (the Institute for Applied Social Sciences) are discussed in detail below.

Interviewer Qualification

Thorough training and qualification activities for interviewers are an essential requirement for imparting the set of standard rules with regard to conducting interviews. These activities primarily aim at providing all interviewers with rules for standardised behaviour in interview situations and committing them to comply with the rules.

In practice, there are three important building blocks in the design of good interviewer qualification:

- Basic training to impart fundamental skills and knowledge
- Project-specific trainings
- Additional subsequent training and debriefing sessions

infas has its own training team available, which is responsible for all training measures at the institute.

Basic Training

Participating in a basic training course is mandatory for all interviewers working in the infas call centre and face-to-face department. The basic training at infas is a one-day event imparting theoretical aspects of standardised interview procedures and providing insights into the tasks of a professional interviewer, as well as the demands and expectations that the role faces.

In addition to theory, the basic training also offers sufficient opportunities for practical exercises by means of the simulation of practical contact or interview situations in a role play. The future interviewers are also instructed in the use of computers and the relevant survey software.

The main focal points of the basic training event are the general features of both contacting and interviewing. The former includes the following issues:

- how to prepare for establishing contact,
- how to communicate throughout the process of contacting and how to avoid refusals, and
- how to document contact results.

For contacting, the primary task is a thorough processing of all addresses from the gross sample. Along with the frequency and variation of an interviewer's contact attempts, the issue of how to handle refusals also plays a major role. Therefore, practical exercises with regard to refusal avoidance are part of the training programme, preparing the interviewer for different contact situations (Shuttles 2012, Groves/McGonagle 2001, Schnell/Beck 2004).

The documentation of contact results is vital for monitoring the interviewers' contacting behaviour. For this purpose, the interviewers use standardised distribution codes[2] to record the result of each single contact attempt during fieldwork. The individual distribution code definitions and their use are explained in the basic training.

The basic training also provides standard rules concerning interviewing:

- Reading out the questions
- Probing
- Data entry
- Handling of specific questionnaire conventions

Interviewing standards provide uniform rules for interviewers when working on the questionnaire. These standard regulations are documented in a basic interviewer manual. At the basic training all interviewers receive this training manually and additional hand-outs.

The participation in basic training at infas ends with a brief test consisting of a set of multiple choice questions regarding the training contents. In addition to checking these contents, the test is also intended to check the participants' extra-functional qualifications. The test protocol is forwarded to the training participants via e-mail, which thus checks the technical aspect of handling infas's usu-

2 Return codes according to AAPOR – The American Association for Public Opinion Research (2011).

al paths of transmission. By completing and returning the test protocol by the due date, the training participants demonstrate their competence and reliability in this regard.

Project-Specific Trainings

Whereas basic training deals with cross-project standards of interviewing and the collaboration between interviewer and survey agency in general, project-specific trainings refer to specific surveys and address the interviewers who are to conduct these particular interviews. Project-specific trainings primarily impart specific survey contents and the demands on the interviews. The focal points of the training are on interviewing and the actual questionnaire. Role-plays and practical exercises rehearse the questionnaire with different interview situations and potential respondent types. With regard to contacting, the interviewers receive instructions on how to use the contact addresses. They also receive aids for argumentation when contacting the specific target groups for this particular project. In project-specific trainings participants expand their experience and increase their knowledge of refusal avoidance techniques, learning how to adapt these to the project-specific conditions.

As a general rule, project-specific trainings also involve the client or research team. This provides them with the opportunity to share their initial experiences directly with the interviewers concerning the use of the instruments. It also enables them to quickly pass on further information to the interviewers where queries arise on the handling of individual questions.

The participation in project-specific trainings is mandatory for all the interviewers who are conducting the interviews for a particular fieldwork project. Depending on a survey's demands and its degree of complexity, these training sessions might last just half a day or several days. The participating interviewers receive the training's contents as a hand-out. An additional comprehensive interviewer manual is also provided to describe the particular interviewer tasks for that project.

Some projects require successful participation in project-specific training to be confirmed by a test interview. In this case, all the interviewers have to conduct a test interview immediately after training and they will not receive their assigned addresses until the training team has approved the test interview.

Debriefing and Subsequent Training

The survey agency's training events are a good opportunity for continuous further qualification of the interviewers. Moreover, additional trainings are measures to fine-tune interviewer behaviour and to deal individually with inappropriate behaviour during fieldwork. Additional trainings generally result from continuous data analysis in the course of a survey and monitoring relevant parameters (see paragraph "Monitoring"). The infas training team is responsible for additional trainings. Measures and consequences for additional trainings are discussed and determined jointly with the project's management. Additional trainings for face-to-face interviewers are usually carried out by telephone to make individual interviewers aware of their errors. In case of frequent interviewer errors, additional training or written instructions can be given to all the project's interviewers or selected groups of them. CATI interviewers receive additional trainings by means of feedback talks immediately after ending the respective interview. As a general rule, all additional interviewer instructions communicated to the interviewers in the course of a project are automatically incorporated into the project's interviewer manual, i.e. the interviewers receive printed sheets with instructions where to insert them in their manual.

In contrast to the additional trainings, the debriefing sessions enable an exchange of experiences between interviewer and research team. Debriefings are generally designed as workshops. They also serve to provide hints during an ongoing project concerning such aspects as possible field strategies. Debriefings after completion of the fieldwork help the research team with the interpretation of the collected data. And debriefings also provide indications of interviewers' actions in the field, their handling of questions or even their approach to conflict situations.

Interviewer motivation

Interviewer motivation aims to provide stimuli for the interviewers to maintain compliance with the interviewer regulations. The primary stimuli are extrinsic, such as interviewers' remuneration. This is especially relevant considering the background that interviewers in Germany normally work freelance at survey agencies. Adequate payment is therefore a substantial part of interviewer motivation and the remuneration model should take into account the efforts of interviewing as well as contacting the addresses. The interviewing effort is based on the duration of each single interview. In the age of computer-assisted interviews (CAI), the actual duration of an interview can be measured by means of the computer's system time. For their contact attempts, the interviewers also receive remuneration based on kilometres driven. Moreover, the remuneration model includes incentives which assess an interviewer's performance via project-specific qualitative and quantitative aspects. The interviewers also receive a daily allowance and travelling expenses for participating in trainings. A basic prerequisite for adequate remuneration also means prompt settlement of the interviewers' accounts, so the survey agency should process and pay all invoices at least once a month.

Along with extrinsic aspects, various intrinsic stimuli can motivate interviewers to work well as an interviewer. The way that the interviewer and survey agency work together is the decisive factor for intrinsic stimuli. Contact with interviewers is intensified through numerous trainings. The interviewers are involved in the project as regards contents and permanently exchange information with the supervisors or field management, as well as the project management or research team. Feedback loops, additional trainings and debriefings support the exchange of experiences between the interviewers and the survey agency. infas regards the interviewers as experts regarding the actual local conditions for conducting interviews.

Monitoring

Continuous monitoring is essential to check the interviewers' compliance with the standards. Monitoring focuses on performance and quality with regard to contacting addresses of the gross sample as well as on the collected survey data. These two aspects are most important for interviewer monitoring. Regarding the contacting of addresses, it has to be guaranteed that the interviewers work only on the given addresses and do not select other target persons on their own. Concerning the collected data, interviewers are monitored to check they comply with standard rules and do not influence the target persons' answers.

Monitoring uses a variety of information and data sources. These include statistics on response rates, cooperation rates, refusals and contact attempts, collected as percentages, quotas and averages and including results from interviewer monitoring via mail or telephone and evaluations derived from observed or recorded interviews.

Parameters

Extensive figures for calculating parameters are available from the distribution codes of individual addresses of the gross sample. The interviewer documents the status for each single address in detail. The survey agency can access them in the form of contact histories and distribution codes to calculate distributions concerning coverage and refusal rates. The evaluation of distributions is based on a comparison of average values (benchmark) for individual interviewers – regarding the entire project or previous survey waves.

During fieldwork, a closer look at the cooperation rate to assess the interviewers' performance has proven to be helpful. The cooperation rate is calculated as the number of complete and partial interviews divided by the sum of the number of interviews (complete plus partial) and the number of non-interviews that involve the identification of and contact with an eligible respondent who refuses (COOP4 – AAPOR 2011, p. 49). An interviewer's success can be assessed by the cooperation rate, independently of the availability of individual respondents or the definition of eligibility. Therefore, the cooperation rate is already meaningful right at the start of the fieldwork. The distribution of individual reasons for refusals is also checked throughout the ongoing fieldwork to identify refusal patterns for individual interviewers.

After fieldwork, a closer look at response rates provides information about the careful processing of all addresses of a sample, which will be used for the final assessment of the interviewers' work. The careful contacting of all addresses should have already been monitored continuously throughout the fieldwork phase. This is an essential part of monitoring in the CAPI field[3], which enables an interviewer's particular contacting strategies or patterns to be detected, in order to avoid selective work on the gross sample. All interviewers receive a weekly report indicating the working progress of the addresses assigned to them. The report includes – along with the number of realised interviews within the week – detailed analyses of time and weekday of the contacts attempted with unavailable respondents, which provide information about the variability of their contacting attempts.

Along with quantitative parameters, numerous clues concerning the interviewers' performance are derived from a content-related analysis of the survey data. Particular attention must be paid to questions with regard to sensitive topics. This is based on the assumption that a certain interviewer partiality can influence the respondent's responsiveness, i.e. the interviewer does not comply with the predefined standard rules. Within this context, the interviewers' quotas

3 The CATI manager – not the interviewer – controls the addresses' use and administration in the CATI field.

for the respondents' willingness to remain in the panel (and thus participate in several waves of the survey) as well as their permission to merge personal data (informed consent to, for example, linkage with register data from social benefit) are compared. The shares of item non-response even show whether interviewers probe correctly or whether they trigger item non-response.

A comparison of the interviewers' average interview duration can provide indications of irregular interviewer behaviour when it comes to handling the questionnaire. Where interview durations are below average, it needs to be checked whether the interviewers indeed read the entire wording of a question and whether they skip questions at certain points. Monitoring telephone interviews allows direct access to this kind of information. For face-to-face fieldwork, the collected data is scanned for hints or indications of deviant interviewer behaviour. In addition, distributions of response concerning selected variables are used to clarify this issue (patterns of response, distribution of prevalence).

Where samples include given addresses and available individual characteristics (e.g. from registry offices), it is also possible to check whether the interviewers complied with the contacting rules and interviewed the correct target person. This is done by comparing individual sample characteristics of respondents with the information provided in the questionnaire.

Interviewer Checks by Mail or by Telephone

Interviewer checks by mail are generally used to check how the interviews were conducted. They are mainly used in the face-to-face field in order to ascertain whether the interviewer actually was on-site and interviewed the predetermined respondent. This type of check can often be skipped for computer-assisted telephone interviews (CATI). As a general rule, telephone interviews are conducted centrally at one call centre where interviewers are under permanent supervision. Furthermore, many call centres use diallers, where the selected telephone numbers are dialled by a computer.

With face-to-face surveys, 20 percent of all interviews conducted are checked by default via postal interviewer checks[4]. infas randomly selects and despatches monitoring questionnaires upon receipt of the realised interviews. Depending on the requirements of the respective survey, however, the monitoring questionnaires can even be dispatched to all respondents and not only a sample. This is now standard procedure at infas for all panel surveys. The send-

4 Different percentages also can be defined for a single survey. The number of checked interviews varies between surveys and depends on the researchers' needs. The proportion of all conducted interviews checked varies from 10 to 25 percent. Schnell (2012, p227)

ing of a thank-you letter to the target person after the study offers the advantage of being able to add the monitoring questionnaire to it. The monitoring questionnaire consists of a set of standardised questions which primarily address the realisation of the interview but also enquire about the respondent's satisfaction with the interview as such and with the interviewer. Eight standard questions in the monitoring questionnaire are used for all surveys and can be extended by including additional project-specific questions. As soon as monitoring samples raise suspicion of potentially faked interviews, a check of 100 percent of the respective interviewer's interviews has to be initiated. Further interviewer checks via telephone are then also conducted.

The survey agency continually records and analyses the returned monitoring questionnaires. The respondents receive a stamped envelope for returning their questionnaire. The recorded data is imported into a database and can thus be assessed for each interviewer as well as each survey. Analysis takes place once a week parallel to a survey's fieldwork and results in a report. Analyses from the interviewer monitoring are also recorded in the interviewer database and the supervising team should then take them into account when planning deployment on further surveys.

Watching and Recording

Watching is a major instrument of supervision in call centres and generally takes place to check the interviewers' compliance with contacting and interviewing standards. Moreover, watching makes it possible to support the interviewers in difficult situations. A watching function within the CATI software enables the supervisors to listen to the conversation between interviewer and respondent and to simultaneously observe the data entry on screen.

In the infas call centre, watching happens in a randomised way. For this purpose, CATI stations are selected at random within predetermined time frames in the course of a day. A supervisor watches the work of an interviewer throughout this time frame, i.e. his/her contacting approach as well as the interview itself. Watching is the basis for the supervisors to assess interviewer behaviour, with the assessment based upon listed criteria of standard rules of interviewing. The respective supervisor discusses his/her watching assessment directly with the interviewer in a brief feedback meeting. The assessments are also included in the interviewer database and the supervising team can take them into account when planning deployment on further surveys.

Recording is possible for both telephone interviews and face-to-face interviews, although face-to-face surveys are recorded less often because an external device (e.g. microphone) is required which might influence the interviewee during the interview situation. When using CAPI, this problem can be overcome by internal laptop microphones. These, however, suffer from the disadvantage of poor sound quality. The respondent's consent is always necessary for any kind of recording. Recordings are used in many surveys to enable the research team to gain insight into interviews without actually being directly involved. They are particularly useful at the beginning of the field period to help identify potential problems the survey instrument may pose with specific target groups. They can also be used for training purposes and feedback meetings with the interviewers.

Reporting

Combining and merging all the monitoring statistics and data while fieldwork is ongoing maximises the transparency of the data collection process shown in the reporting. Professional surveys increasingly make use of specific computer programmes to direct and control fieldwork processes. These fieldwork management systems consist of several connected databases which make it possible to, for example, manage samples, store address data, export mailings or data for fieldwork, and import data from the field. infas has a Sample Management System (iSMS) which is used to control all surveys conducted by the survey agency – independent of their complexity or the survey method. The iSMS administers the gross sample and processes the returned data throughout the ongoing fieldwork. Data and figures are updated daily and exported from iSMS for monitoring purposes. The figures are edited into an extensive standard report so that they provide a basis for detailed analyses and assessment of the field progress. Potential problems or deviations can be quickly identified during the fieldwork phase. Moreover, the figures objectivise communication with all parties involved at the survey agency.

The reporting can result in either individual analyses for each project or a complete overview of all current projects. The status of the gross sample can be accessed via the distributions of the distribution codes. The distributions also serve to document the field's progress for the client. Decisions about possible additional or modified field strategies can be made on the basis of processing status and coverage rates.

The reporting of distributions on the basis of single interviewers is vital for interviewer control and monitoring. Distributions can be considered on a project-specific level as well as on a cross-project level. Clues about an interviewer's potential inappropriate behaviour or performance problems in a project can thus be checked immediately with regard to impact on other projects. The distribution over all interviewers (benchmark) is the primary criterion for interviewer assessment. The figures from reporting serve as the basis for the supervisors' deployment control of their interviewers. For this purpose the interviewers receive weekly reports, as described above.

Regular analysis and evaluation of reporting figures is a major part of data collection within a project.

Summary and Conclusions

Neither deviant behaviour among interviewers nor interviewer errors can be completely avoided in real-life survey practice. Nonetheless, there are a number of strategies and concepts which help to spot errors in time and correct them. Ideally, completely false behaviour among interviewers can be excluded in advance by means of comprehensive training and qualification programmes and sufficient levels of interviewer motivation. Standardised rules form the basic knowledge needed by the interviewers, which is then deepened with practical training and simulation of a variety of different interview situations. Interviewer motivation can be achieved with an appropriate and performance-related remuneration. This should also function as an incentive for the interviewers to comply with the standard rules.

The everyday experience of survey agencies shows that these strategies can eliminate completely erratic behaviour for the most part, but there are still other opportunities for deviation. Interviewers are not necessarily aware of irregularities if they have to make up a practicable solution for a special survey situation on their own. Another source of deviation can be traced back to optimising strategies or falsely learned behaviour among the interviewers. This is where checking and monitoring processes become applicable, both during interview situations and in evaluations of data after it has been collected. An important part of monitoring is a continuous monitoring and evaluation process at the time when the field work is being conducted. This includes not only the initial monitoring and evaluation processes but also the evaluation of (quantitative and qualitative) parameters.

The results have to be made transparent and accessible to all persons participating in the fieldwork process (reporting). Feedback to interviewers and/or consequences for their fieldwork are passed on immediately to avoid more mistakes and stop deviant behaviour in time.

Bibliography

AAPOR (American Association for Public Opinion Research): Standard Definitions: Final Dispositions of Case Codes and Outcome Rates for Surveys, 7th Edition, Lenexa, Kansas, 2011.

Fowler, Floyd J. Jr., Mangione, Thomas W.: Standardized Survey Interviewing, Boston, 1990.

Groves, Robert. M., McGonagle, K. A.: A Theory Guided Interviewing Training Protocol Regarding Survey Participation; in: Journal of Official Statistics, 2001 17, 2, pp. 249-265.

Prüfer, Peter, Stiegler, Angelika: Die Durchführung standardisierter Interviews: Ein Leitfaden, ZUMA How-to-Reihe, Nr. 11, Mannheim, 2002.

Schnell, Rainer., Beck, Christina.: Interviewerverhalten in der Kontaktphase – Information und Ergebnisse der ESS-Schulung zum Kontaktverhalten vom 26.08.2004, unver. Manuskript, 2004.

Schnell, Rainer: Survey Interviews. Methoden standardisierter Befragungen, Wiesbaden 2012.

Shuttles, Charles D.: Refusal Avoidance Training (RAT) in: Lavrakas, Paul J. (ed.): Encyclopaedia of Survey Research Methods. Thousand Oaks: Sage Publications, Inc., null. Sage Research Methods Online, Web, 12 March 2012.

Automatic Interview Control of Market Research Studies

Oliver Hülser

Abstract

Market research studies have to be conducted within very short time frames and should take into account the rising demands on quality standards. Under this postulates it is very important to have an easy-to-use tool to make interview control more efficient. GfK (Gesellschaft für Konsumforschung) developed its own software named ALBERTA. This software includes several tests that systematically check the quality of interviews and interviewers. The "Comprehensive Check of Consistency" and the "Interview Variance Analysis" are tests on exceptional patterns of answers in the whole questionnaire. The "Internal Check of Consistency" analyses the variance in statement batteries, and the "Identification of Doubled Interviews" detects very similar interviews. By means of statistical criteria ALBERTA systematically identifies conspicuous interviews and interviewers. Researchers however, are responsible for carefully deciding which interviews are to be deleted and can thus reduce the risk of eliminating false positives.

Challenges to Market Research

Many decisions in the daily business of companies are based on the results of market research. To a large extent the results of market research are built on interviews with consumers. Therefore, the quality of interviews is the basis for the quality of market research. Many different survey methods like telephone interviews, web interviews, and personal interviews with or without laptop support are used. There are also many international studies involving interviews in different countries and several field institutes. A huge amount of interviews have to be realized in many different studies that are usually facing a very strict time limit. In 2010, almost two million interviews were conducted in Germany in the field of GfK ad hoc research (Figure 6.1). Ad hoc research at GfK includes all quantitative surveys without panels. The quality of an interview largely depends on the respondent. It is very important that a respondent gives honest answers and that he or she completes the whole interview. One quarter of the two million interviews were conducted by 3300 interviewers. The quality of those interviews depends to a large extend on the good work of the interviewers. It is highly preferable to identify interviews of poor quality, no matter whether they are of poor quality due to the interviewer or the respondent.

To develop strategies to identify false interviews it is helpful to think of possible reasons why interviews are faked, and which conditions lead to this.

The participation in market research studies is voluntary. Therefore, it seems unlikely that a respondent who takes part by choice does not give correct answers. However, several reasons for not giving proper answers in an interview can be identified: The interview is too long and too boring. Single questions are difficult to understand or are missing a "don't know" option. Delicate questions are not answered honestly. The respondent only participates to get an incentive.

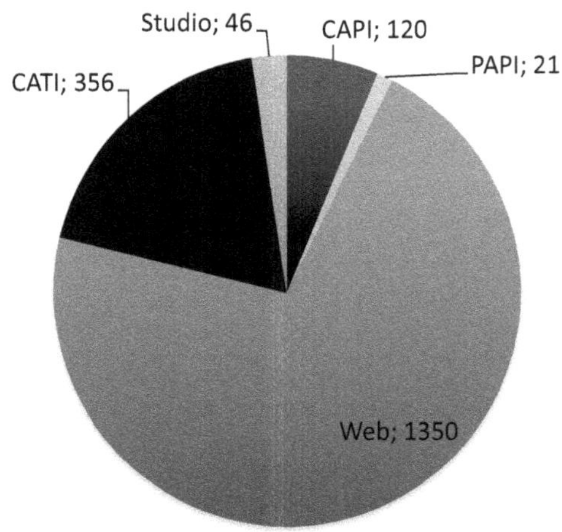

Fig. 6.1: Interviews (in 1000) in GfK Ad hoc Research in Germany 2010

When doing quality checks the concept of a particular study has to be taken into account. A questionnaire that is too long can lead to bad answers in the last part of the interviews. Hence respondents try to finish the interview as fast as possible. Respondents might refuse to answer delicate questions for certain reasons and rather give wrong than no answers. These include questions regarding income, drug consumption, or sexual habits.

The risk to receive false interviews is higher if the study is carried out by interviewers. Interviewers may have multiple reasons to fake interviews or at least parts of interviews. The rate of actual faked interviews will differ strongly and the real amount can hardly be determined. It is estimated that up to about five percent of the interviews conducted by interviewers are fakes (Biemer and Stokes 1989; Schnell 1991). A large amount of those fakes are not entire but partial fakes and thus much more difficult to identify. Parts of those interviews are correct, because the interviewer at least contacted the respondent and asked a few questions (Reuband 1990; Biemer and Stokes 1989; Blasius and Thiessen 2012). A partially faked interview is sometimes not even identified when the respondent is contacted a second time by a different interviewer. It is possible that respondents remember only that the interview took place, but they forgot the length and the number of questions. A partial faked interview that is identified has to be treated the same way as a complete fake. It has to be deleted.

The more interviews an interviewer realizes in a certain amount of time, the more money he or she earns. For the interviewer it is time-consuming to seek for certain target groups with special quota criteria, and it takes costs and time for travelling to realize interviews.

This shows that there are plenty of reasons to fake interviews. Therefore, quality checks have to deal with many different kinds of undesirable effects. It is possible that only parts of a questionnaire are faked or that only quota criteria are not correct, because a wrong target person was interviewed. Entire interviews may be filled out randomly by the interviewer. Furthermore, interviews may be filled out carefully by the interviewer without interviewing anyone, which is much harder to detect. Interviewers may fake all of their interviews or only a few interviews.[1] The less systematic the occurrence of faking interviews is, the more difficult it is to identify those interviews.

A Software for an Automatic Check of Interviews

The process of quality control is quite complex in market research. The goal is to check as many interviews as possible in a very short time. Thus, it is obvious that the amount of manual checks is limited and that automated tools are needed to speed up this process. GfK has developed its own tool, named ALBERTA,[2] to detect faked interviews. This software contains several tests to identify suspicious interviews, and systematically helps to find interviews that should be controlled manually. These manual checks can be either a systematic and substantial check of the data itself or a re-interview. The re-interview helps to find out if the interview took place with the correct person to an adequate extent. This is probably the most effective method (Biemer and Stokes 1989; Feindt et al 1997), but it is also the most expensive and time consuming method, and it can be done only for selected interviews.

The basic principle of ALBERTA is to find either interviewers who are systematically cheating or to find single interviews that are of poor quality. This is done by using only statistical criteria. No hypotheses with regards to the content

1 It is very difficult to do research with real fakers, because usually they do not cooperate. In the course of the IFiS project (Identification of Falsifications in Survey Data) Güllner and Porst (2012) conducted a qualitative study on strategies while falsifying interview data.

2 The initial concept for ALBERTA was developed by Raimund Wildner.

are needed to check the plausibility of data. The result of an ALBERTA analysis is a list of interviews and/or interviewers that might be of poor quality. The researcher has to decide whether poor interviews are deleted or not.

This tool effectively improves the quality of surveys and reduces the input for an intensive unassembled examination of interviews. It makes the data check much more efficient and helps to find implausible cases in the data before the statistical analyses. Last but not least, it supports the supervision of interviewers.

Many indicators to identify a faked interview are based on the deviation from the average of other interviews. Nevertheless, an interview cannot be judged as a suspicious interview because of one single indicator. Several indicators have to draw a clear picture whether an interview is a potential fake.[3] The risk to sort out false positives, which is the elimination of interviews that are identified as fakes but actually are correct, is high and has to be taken into account. An interview should not be flagged as a fake with levity. If scales with extreme answers are provided in a questionnaire, it is in principle allowed to give those answers. Respondents with extreme but consistent answers should not be sorted out.

The criteria to flag suspicious interviews may vary, because the quality of studies is varying as well. Many suspicious interviews in a study are not necessarily an indicator for bad respondents or interviewers, but may be a hint that a study is too long or too boring. ALBERTA not only identifies bad interviews in single studies but is also part of the whole process to judge interviewers and studies in the long run.

ALBERTA has been in use for several years at GfK in many countries. It is standalone software that works with SPSS data. The current version provides four tests to check the quality of interviews and interviewers:

1. Comprehensive Check of Consistency
2. Interview Variance Analysis
3. Internal Check of Consistency
4. Identification of Doubled Interviews

The Comprehensive Check of Consistency and the Interview Variance Analysis are tests on exceptional patterns of answers in the whole questionnaire. The Internal Check of Consistency tests the variance in statement batteries and the Identification of Doubled Interviews seeks for very similar interviews.

---

3 ALBERTA gives a descriptive summary of the different indicators and the researcher has to decide, which combination of indicators might be a potential fake. Bredl et al. 2012 use cluster analysis and discriminant analysis to identify fakers out of a set of indicators.

Comprehensive Check of Consistency

This first test is based on the idea that respondents who do not answer a questionnaire with care give answers that are inconsistent. Combinations of inconsistent answers are much more unlikely than combinations of consistent answers. The test executes a theory free examination of all variables with less than twenty different values. To compute this test, each variable is split up into dummies, which leads to a great amount of variables. Every dummy is chi^2 tested with every other dummy and for all highly significant statistical relations punishment scores are calculated. As a consequence, improbable responses get more punishment scores than probable responses. The plausibility of interviews with very high punishment scores have to be examined. The check of the interviewer has no priority in this test.

The example in Table 6.1 shows the cross tabulation of any two dummies. The chi^2 value is 13.5 and thus highly significant at the 99.9 level. For those two dummies ALBERTA calculates the punishment scores shown in Table 6.2.

Table 6.1: Example Comprehensive Check of Consistency

	Dummy 2		
Dummy 1	0	1	Overall
0	37	43	80
1	18	2	20
Overall	55	45	100

Table 6.2: Punishment Scores Comprehensive Check of Consistency

	Dummy 2	
Dummy 1	0	1
0	1.17	0.67
1	1.22	5.42

In this case the most unlikely combination of answers is a 1 in both dummies. The punishment score for this combination is by more than four times higher than for all other combinations. This does not mean that the two interviews that contain this combination 1/1 are faked interviews. They are flagged as probably faked interviews if they have many more strange combinations, and in the end the sum of their punishment scores is remarkably high.

Interview Variance Analysis

The central hypothesis of the Interview Variance Analysis is that interviewers who fake complete interviews produce less variance than real respondents.

The test is based on the method described by Schäfer et al. (2004), who named the test Variability Method. The authors identified some reasons why interviewers who fake interviews produce less variance than honest interviewers. Fakers tend to answer all questions and produce therefore fewer missings. Moreover, fakers do not misunderstand the questionnaire and avoid extreme or conspicuous answers (Schäfer et al. 2004). For example they tend to give rather socially desirable than controversial answers. The test is based on the comparison of variance of interviewers: the lower the variance of an interviewer the more likely it is that this interviewer is a faker.

To make the test applicable for all different kinds of market research data and questionnaires containing very different scale levels the test is adapted in ALBERTA. The test uses 20 percent of all variables of a questionnaire that have the highest diversification.[4] In order to make the individual answers comparable, a z-transformation is undertaken for each variable. Thus every variable used in the test has a mean of 0 with a variance of 1.

Those selected and standardized variables are used to compute the standard deviation per interview and - as shown - the average standard deviation per interviewer:

$$V_l = \frac{1}{g}\sum_{k=1}^{g}\sqrt{\frac{1}{f-1}\sum_{j=1}^{f}[Z_{ij}-\bar{Z}_i]^2}$$

with:

V_l = Standard deviation of the z-transformed variables f of interviewer l
g = Number of interviews (k=1,2,...,g) per interviewer l
f = Number of included variables (j=1,2,...,f)
$\bar{Z}_i$ = Mean z-score of respondent i of all included variables
Z_{ij}= The interval on question j in interview i, which has been submitted to a z- transformation

Interviews and interviewers with a variance that is two times lower than the average variance are marked.

4 A minimum of 100 variables is used if the data file consists of less than 500 variables.

In analogy to the Comprehensive Check of Consistency interviews having a very large variance might be also conspicuous and are therefore marked as well. In this case the test refers to respondents rather than to interviewers.

Additionally, the percentage of missings per interview is shown in the output table and is also used as an indicator.

Internal Check of Consistency

The usage of item batteries with rating scales is very common in market research. There is a high risk especially in studies without interviewers (e.g. online studies) that respondents do not read the questions carefully. Relating to the number of interviews these are mostly online studies. This may lead to some respondents either filling out statement batteries randomly or straight lining them. If respondents fill out the item batteries randomly, the variance within this statement battery is extraordinarily high. The variance of statement batteries that are partly or completely straight lined is very low. The effect of (partly) straight lining occurs also in studies, which were conducted by interviewers.

The internal check of consistency is done by a comparison of the standard deviations (SD) in the replies of item blocks. Two control numbers, one for a strong diversification and the other for a low diversification, are calculated for each interview. This is done by summing up the number of item blocks, where the individual SD is remarkably greater or lower than the average SD.

The SD of the statements (T_i) is then calculated per item block and interview:

$$T_i = \sqrt{\frac{1}{k-1}\sum_{j=1}^{k}[X_{ij} - \bar{X}_i]^2}$$

with:

T_i = SD of statements of respondent i
k = Number of items in the block (j=1,2,...,k)
i = Respondent
X_{ij}= Value of respondent i for item j
$\bar{X}_i$= Mean value of the respondent i throughout all k items

Based on the individual SD the average SD of this item block for all respondents is calculated and finally the 'SD of the SD' (TT) is calculated per item block:

$$T_T = \sqrt{\frac{1}{n-1}\sum_{i=1}^{n}\left\{\sqrt{\frac{1}{k-1}\sum_{j=1}^{k}[X_{ij}-\bar{X}_i]^2}-\frac{1}{n}\sum_{i=1}^{n}\sqrt{\frac{1}{k-1}\sum_{j=1}^{k}[X_{ij}-\bar{X}_i]^2}\right\}^2}$$

with:

T_T = SD of the individual SD of T_i
n = Number of respondents (i=1,2,...,n)
k = Number of items in the block (j=1,2,...,k)
X_{ij}= Value of respondent i for item j
$\bar{X}_i$= Mean value of the respondent i throughout all k items

The control number for a strong diversification is the sum of item blocks with an individual SD (Ti) greater than the average SD plus twice the 'SD of the SD' (TT). Analogically, the control number for a low diversification is the sum of item blocks with the individual SD (Ti) smaller than the average SD minus twice the 'SD of the SD' (TT). If the standard deviation is zero, the control number for a low diversification is raised by one, even if the average diversification is not remarkably different. The interviews with the highest values in those two control numbers will be checked.

The control number for a strong diversification is the sum of item blocks where the individual SD (Ti) is two times greater than the 'SD of the SD' (TT) upwards to the average SD of an item block. Analogically, the control number for a low diversification is the sum of item blocks where the individual SD (Ti) differs less than two 'SD of the SD' (TT) to the average SD of an item block. If the standard deviation is zero, the control number for a low diversification is raised by one, even if the average diversification is not remarkably different. The interviews with the highest values in those two control numbers will be checked.

An additional outcome of this test is the evaluation of surveys. Many interviews with poor results in the internal check of consistency do not necessarily judge the respondents or the interviewers. In fact, this is an indication on the quality of the survey respectively the questionnaire. Item blocks with many respondents having no variance can be taken out of the questionnaire or changed in further studies.

Identification of Doubled Interviews

This final test is based on the initial hypothesis that deliberately doubled interviews will at least differ in details. Interviewers who double interviews will at least change a few variables to hide their foul play or change demographics to meet quota criteria. This means that slight differences in doubled interviews are possible and quite likely. Conventional tests that check for complete equality of interviews fail in this case. Checks on identical responses (for example Blasius and Thiessen 2012) require more efforts to prepare the data because only parts are checked. ALBERTA is able to handle all variables of a study. The researcher has no need to think about reasonable variables for the test on doubled interviews.

To reduce the complexity and thus the computing time, this test is based on the same variables with the highest diversification as the Interview Variance Test. Based on these variables the distance between every interview and every other interview is computed by summing up the squared Euclidian distances. The most similar cases have to be checked. In some studies it is difficult to judge whether an interviewer doubles interviews or respondents are likely to give similar answers. In this case it is important to compare the similarity of interviews of one interviewer with similar interviews of different interviewers.

Case Study

In 2011 GfK conducted the study "Who is Germany" for the Leibniz Institute for the Social Sciences GESIS. For this study 81 interviewers did 1200 face-to-face interviews of a questionnaire with 300 variables and 12 statement batteries.[5] For the interview check all four tests of ALBERTA were used.

Comprehensive Check of Consistency: Results

The punishment scores of the Comprehensive Check of Consistency are normalized in a way that the maximum score is 10000. The Comprehensive Check of Consistency identified two very suspicious interviews as well as seven more interviews with punishment scores greater than 8000 that are probably conspicuous. A closer look on those interviews did not confirm that they were faked interviews.

5 First results of this study are presented in Kemper et al. 2012.

A common way to visualize such results is a scree plot that shows how different the most suspicious cases are. To make the scree plot more clearly only the 100 cases with the highest punishment scores are shown in Figure 6.2.

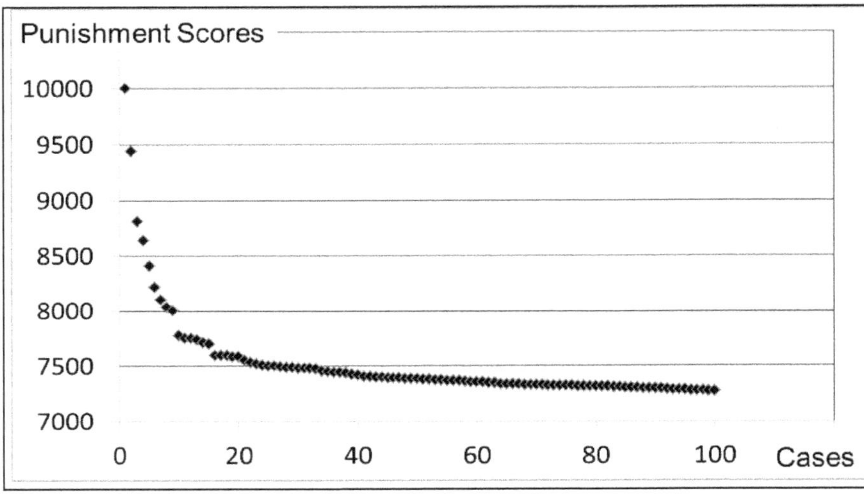

Fig. 6.2: *Punishment scores of the Comprehensive Check of Consistency*

Interview Variance Analysis: Results

The Interview Variance Test marks three interviewers as suspicious because their average interview variance is remarkably low (table 6.3).[6] The most suspicious interviewer (559) conducted only one interview, therefore it is not possible to judge this interviewer. Three out of five interviews of interviewer 356 have a very low interview variance, which leads to very low average variance of this interviewer. The interviews of this interviewer are likely to be fakes. To confirm this, re-interviewing would be the best option. Interviewer 339 who conducted 18 interviews is very suspicious. All of his or her interviews have an interview variance that is clearly below the average variance of .98. It will be shown in a later section that those interviews are identified as doubled as well. It is very likely that this interviewer faked most if not all of his interviews. The last interviewer (98) shown in Table 6.3 did 14 interviews. The average variance of this

6 If the average interview variance of an interviewer is remarkably low, all interviews of this interviewer are marked as suspicious irrespective of the specific interview variance. If the average interview variance of an interviewer is not remarkable low, only single interviews with low variance can be marked as suspicious.

interviewer is not remarkable low. Only one of his or her interviews is suspicious because of low variance. However, based on only this indicator the interviewer cannot be judged as a faker.

Table 6.3: Interview Variance Test

Interviewer-ID (Number of Interviews)	Interview-ID	Suspicous	Interview Variance per Interview (average=.98)	Average Interview Variance per Interviewer
559 (1)	70	X	.695	.695
356 (5)	189	X	.633	.745
356 (5)	187	X	.652	.745
356 (5)	188	X	.684	.745
356 (5)	190	X	.847	.745
356 (5)	186	X	.908	.745
339 (18)	43	X	.679	.766
339 (18)	46	X	.693	.766
339 (18)	54	X	.718	.766
339 (18)	42	X	.738	.766
339 (18)	52	X	.74	.766
339 (18)	48	X	.741	.766
339 (18)	50	X	.75	.766
339 (18)	106	X	.754	.766
339 (18)	44	X	.758	.766
339 (18)	53	X	.759	.766
339 (18)	10	X	.761	.766
339 (18)	47	X	.765	.766
339 (18)	45	X	.781	.766
339 (18)	105	X	.805	.766
339 (18)	51	X	.819	.766
339 (18)	1018	X	.821	.766
339 (18)	1017	X	.844	.766
339 (18)	1019	X	.868	.766
98 (14)	835	X	.627	.796
98 (14)	639		.703	.796
98 (14)	636		.733	.796

Internal Check of Consistency: Results

The Internal Check of Consistency on twelve statement batteries identified four interviews with six or more suspicious statement batteries. The extract of the results in Figure 6.3 shows that interview 479 has nine suspect batteries (C4). One battery is completely missing (C3) and eight batteries are suspect because of low variance (C2). Respondent 343 refused to answer seven statement batteries (C3). With almost 44 percent missings in the whole interview, this case seems to be of poor quality anyway. Interview 900 is ambiguous, with four out of twelve batteries with a low variance (C2), one completely missing (C3), and one with a strong variance (C1). The fact that there are only four out of twelve batteries with low variance is only a weak indicator for this interviewer to be a faker. Therefore, the risk to sort out a false positive would be high.

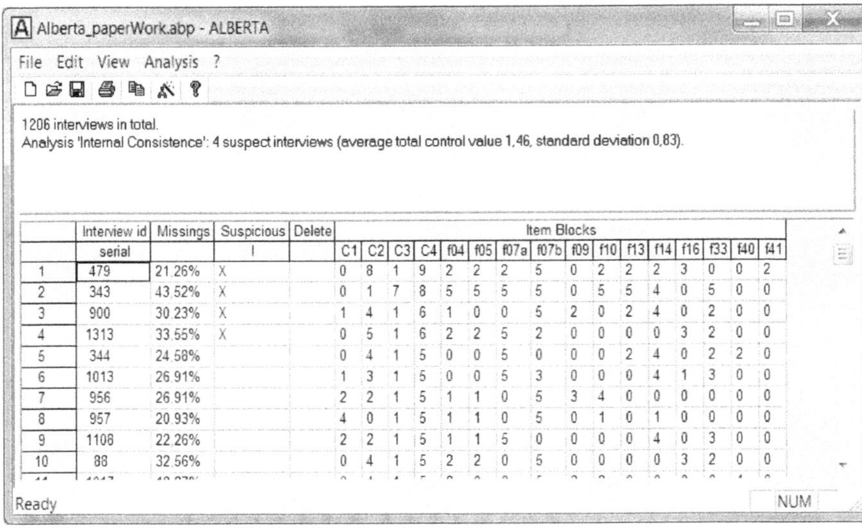

	Interview id	Missings	Suspicious	Delete	Item Blocks															
	serial		I		C1	C2	C3	C4	f04	f05	f07a	f07b	f09	f10	f13	f14	f16	f33	f40	f41
1	479	21,26%	X		0	8	1	9	2	2	2	5	0	2	2	2	3	0	0	2
2	343	43,52%	X		0	1	7	8	5	5	5	5	0	5	5	4	0	5	0	0
3	900	30,23%	X		1	4	1	6	1	0	0	5	2	0	2	4	0	2	0	0
4	1313	33,55%	X		0	5	1	6	2	2	5	2	0	0	0	0	3	2	0	0
5	344	24,58%			0	4	1	5	0	0	5	0	0	0	2	4	0	2	2	0
6	1013	26,91%			1	3	1	5	0	0	5	3	0	0	0	4	1	3	0	0
7	956	26,91%			2	2	1	5	1	1	0	5	3	4	0	0	0	0	0	0
8	957	20,93%			4	0	1	5	1	1	0	5	0	1	0	1	0	0	0	0
9	1108	22,26%			2	2	1	5	1	1	5	0	0	0	0	4	0	3	0	0
10	88	32,56%			0	4	1	5	2	2	0	5	0	0	0	0	3	2	0	0

Fig. 6.3: ALBERTA Result of Internal Check of Consistency

Identification of Doubled Interviews: Results

The test on doubled interviews revealed that the 24 interviews with the most similar partners include all 18 interviews conducted by interviewer 339. This interviewer is already known from the Variance Test. All interviews from this interviewer are not used for any further analysis. Other interviewers did not have any systematic pairs of interviews that led to the conclusion that they would be potential fakers.

Summary Case Study

In summary, the data check in this case study based on real data identified two percent of the interviews as fakes, which is not extraordinary high.[7] The results yield a clear picture for one interviewer. By combining the results of all tests, a clear indication can be seen that the interviews conducted by this interviewer are likely to be fakes. However, only one out of 81 interviewers is labeled as suspicious so that all in all the quality of the data collection of this study is good.

Conclusion

The need for a systematic check of interviews in market research is obvious. The amount of interviews is huge and the demands on quality are high. GfK developed a tool named ALBERTA, which systematically supports the quality control of interviewers and interviews. ALBERTA enables an automated check of interviews and interviewers in an easy-to-use environment at reasonably low costs. It is very effective for identifying obscure interviews and it supports the choice of interviews that have to be checked manually. The work of interviewers can be judged in the long-run and interviewers with a poor quality of work can get trained. An evaluation of study quality is also possible.

The indicators used in ALBERTA are very suitable to control studies in a reasonable time period. However, the work on ALBERTA is always in progress and GfK tries to optimize its tool. Other indicators may be included in the software. Especially the linkage of the indicators may be an improvement.[8]

By means of statistical criteria ALBERTA systematically identifies conspicuous interviews and interviewers. One has to keep in mind that the software does not delete interviews. The researcher is responsible to decide carefully, which interviews are to be deleted. Extraordinary interviews may derive from some kind of flamboyant characters that are authentic and should not be sorted out because they are false positives.

---

7 It strongly depends on the type of study how many faked interviews are expected. A general accepted range does not exist. Biemer and Stokes (1989) and Schnell (1991) expect up to about five percent of the interviews conducted by interviewers are fakes.

8 See Bredl et al. 2012, Storfinger and Winker 2011, Storfinger and Opper 2011 for some possible ideas on indicators to extend ALBERTA.

Bibliography

Biemer, Paul P. and S. Lynne Stokes: The Optimal Design of Quality Control Samples to Detect Interviewer Cheating. Journal of Official Statistics, Vol. 5. No. 1, pp. 23-39. 1989.

Blasius, Jörg and Victor Thiessen 2012: Assessing the Quality of Survey Data. Sage Publications Ltd. 2012.

Bredl, Sebastian, Winker, Peter, Kötschau, Kerstin: A statistical approach to detect interviewer falsification of survey data, Survey Methodology (forthcoming), 2012.

Feindt, Patricia; Irwin Schreiner and John Bushery: Reinterview: a Tool for Survey Quality Improvement. In Proceedings of the Survey Research Methods Section, American Statistical Association, pp. 105-110. 1997.

Güllner, Gesine and Rolf Porst: Identifikation von Fälschungen in Surveys. Bericht über den kognitiven Test im Rahmen des Projekts IFiS. GESIS Working Papers 2012/16. 2012.

Kemper, Christoph J.; Constanze Beierlein; Anastassyia Kovaleva and Beatrice Rammstedt: Eine Kurzskala zur Messung von Optimismus-Pessimismus: Die Skala Optimismus-Pessimismus-2 (SOP2) (GESIS Working Papers 2012|15). Köln: GESIS. 2012.

Reuband, Karl-Heinz: Interviews, die keine sind. „Erfolge“ und „Mißerfolge“ beim Fälschen von Interviews. In: Kölner Zeitschrift für Soziologie und Sozialpsychologie, Jg. 42, Heft 4, S. 706-733. pp. 279-301 Wiley. Chichester. 1990.

Schäfer, Christin; Jörg-Peter Schräpler; Klaus-Robert Müller and Gert G. Wagner: Identification of Faked and Fraudulent Interviews in Surveys by Two Different Methods. DIW Discussion Papers 441: Berlin. 2004.

Schnell, Reiner: Der Einfluss gefälschter Interviews auf Survey-Ergebnisse. Zeitschrift für Soziologie, Jg. 20, Heft 1, pp. 25-35. 1991.

Storfinger, Nina and Peter Winker: Robustness of Clustering Methods for Identification of Potential Falsifications in Survey Data. Discussion Paper No. 57. Center for International Development and Environmental Research (ZEU). Gießen. 2011.

Storfinger, Nina and Marie Opper: Datenbasierte Indikatoren für potenziell abweichendes Interviewerverhalten. Discussion Paper No. 58. Center for International Development and Environmental Research (ZEU). Gießen. 2011.

Controlling Interviewers' Work in PIAAC – the Programme for the International Assessment of Adult Competencies

Natascha Massing, Daniela Ackermann, Silke Martin, Anouk Zabal and Beatrice Rammstedt

Abstract

PIAAC is an international comparative survey initiated by the OECD (Organization for Economic Co-operation and Development). Its goal is the assessment of adult skills and competencies to develop an empirically-based foundation for political interventions. PIAAC is a cross-sectional survey that is being conducted in 25 countries from all over the world. To ensure that the survey data is comparable across countries, the International Consortium of PIAAC has defined specific and extensive quality standards[1]. Each participating country had to comply with these standards and adapt them to the national context as quality control mechanisms. In this chapter, the international quality standards related to interviewer control and their implementation in Germany will be presented. In addition, we will describe the interviewer trainings conducted in Germany and discuss the potential of such trainings in preventing interview falsification.

Introduction

Data quality is essential for surveys: It is only possible to draw valid conclusions from the data collected if errors are minimized and quality is ensured at all stages of data collection. Interviewer control plays an important part in ensuring survey data quality. The major goals of controlling interviewers' work are to detect falsifications and to identify interviewer errors such as interviews conducted with others than the sampled person. Both have a significant impact on data quality and should be eliminated or minimized.

The Programme for the International Assessment of Adult Competencies (PIAAC) is a complex and politically relevant international study. It aims at generating solid empirical evidence regarding fundamental cognitive competencies of adults. Extensive quality control mechanisms have been put in place to achieve the highest possible data quality as well as comparability of results across all countries participating in PIAAC.

The present chapter describes the key standard procedures for controlling and validating interviewers' work in PIAAC. First of all, we will give a brief introduction of PIAAC and its main goals. Next, we will focus on interviewer

---

1 PIAAC Technical Standards and Guidelines, December 2011 (OECD 2011). The only version that is publically accessible dates from October 2008 (OECD 2008).

control as it was implemented in PIAAC: After a review of the international quality control procedures, the national control measures implemented in the German context will be specified. Beyond standard interview control procedures that are widely used in many surveys, there were specific features implemented for PIAAC which are of special interest and are described in this chapter.

Finally, we will turn to the PIAAC interviewer trainings in Germany. This will be done in a separate section as the trainings are not a direct control measure. We will discuss their role as a quality enhancing instrument prior to data collection and consider their role in preventing interview falsifications.

Overview of PIAAC

PIAAC was initiated by the OECD (Organization for Economic Co-operation and Development) and is an international comparative study of adult competencies conducted in 25 countries across the world. In most countries, including Germany, the data collection took place between summer 2011 and spring 2012. One of the central goals of PIAAC is to determine the current distribution of adult skills and competencies and thus, to provide empirical evidence to inform policy-makers. An international Consortium has been given the overall responsibility for the international project management by the OECD. In Germany, the survey is funded by the Federal Ministry of Education and Research with the involvement of the Federal Ministry of Labour and Social Affairs. A team of researchers at GESIS – Leibniz Institute for the Social Sciences is responsible for the national project management in Germany.

In all participating countries, PIAAC was based on random samples of the country's adult population (16-65 yrs.) with a target size of approximately N = 5,000 respondents (net sample). The interview contained two major parts: the background questionnaire and the assessment of competencies. The background questionnaire included questions on the respondent's demographics, education and work as well as the use of skills at work and in every-day life. The assessment included tasks measuring every day skills, assessing adults' literacy and numeracy competencies and their ability to solve problems in technology-rich environments. The interview was designed to have an average duration of 1:45h and was completely computer-assisted. As in many other surveys, CAPI[2] was used for the administration of the background questionnaire. Unlike previous adult literacy surveys which used a paper-based skills assessment, PIAAC was the first to also implement a computer-based assessment of the cognitive instruments. The assessment was self-administered under the supervision of the inter-

2 CAPI stands for Computer-Assisted Personal Interview.

viewer. The majority of the respondents completed the computer-based assessment. However, if a respondent had no or only insufficient computer experience, the tasks could also be completed on paper.

Controlling Interviewers' Work in PIAAC

International Standards and Guidelines Related to Quality Control

> "A large-scale survey is a complex system of interrelated activities and processes that must be well-integrated and controlled in order to produce a quality product" (Lyberg and Biemer 2008 p. 421).

To ensure high-quality data for PIAAC, specific and comprehensive *Technical Standards and Guidelines* were defined by the Consortium.[3] All participating countries had to comply with these standards when carrying out PIAAC. The Consortium monitored the implementation of these standards very closely in each country. If a country did not sufficiently comply with these guidelines, its data might be excluded from the international PIAAC report.

In this chapter we do not look at all aspects of quality control but focus on the most important aspects of the *Standards and Guidelines* concerning interviewer control.

Interviewers have the first personal contact with sampled persons. It is their task to explain the goal of the study and convince sampled persons to participate. Furthermore, they are required to administer the survey instrument in the correct way, including asking survey questions, probing, providing further information if needed, and coding answers into the provided scheme (Schaeffer et al. 2010). In PIAAC, the interviewers' tasks were even more complex: They did not only have to administer the background questionnaire but also the skills' assessment, which entailed additional tasks. It was therefore crucial that interviewers working on the project were proficient in their job and well-trained specifically for PIAAC.

Because interviewers are the first in line when it comes to the data collection process, quality control relating to the data collection has to start with their work. The two most important general PIAAC standards and guidelines relating to interviewer control specified that the performance of interviewers had to be monitored during the field period and that interviewers had to be aware of the

3 The Technical Standards and Guidelines were specified in more detail in a comprehensive Quality Control Manual developed by the Consortium. This document is not publicly available.

quality control mechanisms that were implemented in the field. In addition, some specific standards were defined regarding interviewer validation. One essential requirement was the validation "of 10% of each interviewer's finalised work, including cases finalised as nonresponse" (OECD 2008 p.137). The standard therefore refers to the validation of all finalized cases, i.e. interviews, refusals, non-contacts, disabilities, ineligibles and cases with invalid addresses. The Consortium suggested that these cases be selected randomly and from all final disposition codes for validation.[4]

Beyond that, at least two interviews per interviewer had to be tape recorded early in fieldwork. These tape recordings had to be reviewed in order to control the performance of the interviewer and – in case needed – to provide him or her with feedback or additional training. In addition, the Consortium recommended that countries perform quality checks on open questions, e.g. the questions on occupation which were subsequently used for coding into the International Standard Classification of Occupations (ISCO).

By closely monitoring interviewers' work, the national teams in each country were able to observe the progress of the fieldwork and detect any problems that occurred during the field period. The Consortium recommended looking at the overall pattern of an interviewer's performance rather than verifying single aspects of interviewers' work. A standard stated that if during validation, work by an interviewer would be found to be suspect, 100% of his or her cases needed to be verified.

As mentioned earlier, the implementation of the requested quality control procedures were very closely monitored by the Consortium via different kinds of quality control reports which countries were required to submit throughout the fieldwork period: a sample monitoring form documenting the development of response rates and sample yields submitted on a regular basis, as well as a monthly survey operations quality control form. This latter form covered questions on several aspects of fieldwork staffing, activities and production and was supplemented by quality control calls every two months.

Interviewer Validation and Interview Quality Control: Implementation in Germany

Not all features of the *Standards and Guidelines* were equally suitable for each country. However, they had to be implemented in the best possible way in each

---

4 A disposition code determines the final result of a case and indicates whether the case was a successful interview or another disposition. For PIAAC, interviewers could assign one out of more than 20 dispositions to each case.

country, accounting for national constraints and circumstances. Consequently, in some countries certain standards needed to be adapted. All such adaptations were monitored by the Consortium to ensure that the same quality standards were met in all participating countries. This section specifies the procedures for interviewer validation for PIAAC in Germany.

The survey organization contracted for Germany already had well-established procedures of quality control checks for monitoring interviewers' work. Thus, some of the procedures specified in the *Standards and Guidelines* were part of the regular quality control processes at the survey organization. Others, however, were specifically added for PIAAC. Furthermore, there were some specific topics where no quality control was foreseen internationally, but that were considered to be important so that additional control mechanisms were implemented nationally. Each implemented feature of the quality control procedure was discussed between the survey organization and the GESIS-Team. Most of these procedures were carried out by the survey organization and were controlled by the GESIS-Team.

As for many surveys, monitoring procedures were implemented in Germany to validate interviewers' work during the PIAAC field period. Interviewers were made aware of the quality control procedures during the interviewer trainings. Consequently the two general standards for PIAAC were met. In addition, the following specific monitoring procedures were observed and will be described in detail in the following sections: 1) 10% validation of each interviewer's finalized cases, including non-response, 2) review of tape recordings, 3) look at the overall pattern of an interviewer's performance, 4) validation of 100% of the cases for interviewers with suspect cases, and 5) additional control mechanisms in Germany.

- Validation of 10% of Each Interviewer's Finalized Cases, Including Non-response

In order to achieve the required 10% validation of all finalized cases, in Germany we focused on the validation of completed interviews. As the chances for falsification are the highest with completed interviews, the control of completed cases was considered to be most important. For this reason, we attempted to validate 100% of completed interviews. The two main methods in place to validate PIAAC interviews in Germany used data about the respondent from other sources than the interview: from a) the registry and b) a control questionnaire sent to respondents after completion of the interview as well as to certain non-respondents.

However, as the standards required to also validate non-response, additional checks were implemented, in order to attempt to validate as many other disposition codes as possible.

Controlling Interviews

a) For PIAAC in Germany, a registry-based probability sample was drawn. The sample frame provided information on age, gender and nationality for each respondent – information that was not available for countries using household samples (Koch 1995). Making use of this advantage, the first step of quality control in Germany was to compare the data of all interviews with the data from the registry. The information on age, gender and nationality provided by the registry were systematically compared with the information recorded in the interview. Differences between both could be caused by various reasons, such as wrong registry data, an interviewer typo, incorrect information given by the respondent, the wrong person had been interviewed, or it may have been an indication of a falsification.
b) Another important method of quality control is to re-contact the respondent after the interviewer has finalized the case (ESS 2011). In Germany, each PIAAC respondent[5] received a short standardized control questionnaire to validate the interviewers' work. The control questionnaire was based on the master questionnaire provided by the Consortium with national adaptations, and contained questions on whether the interview had actually taken place, on the respondent's age, gender and education and on interview timing information. The data from this control questionnaire was then compared to the information from the registry as well as to the data recorded in the interview.

If any suspicious differences were identified with one of the methods described above, the whole interview was checked in detail. Also direct contact with the respondent via telephone was attempted in order to investigate and explain the observed inconsistencies. If the respondent could not be contacted the interviewer was asked to explain the differences. If the inconsistency occurred because e.g. the wrong person was interviewed, the data was deleted from the data set and the case was released again.

5 By respondents we mean sampled persons who have participated in PIAAC. One reason to focus on respondents only was that in Germany it is legally not allowed to re-contact all sampled persons without interviews, such as hard refusals, due to data protection laws.

Controlling other Disposition Codes

The control questionnaire sent out to all respondents yielded already the requested validation of over 10% of the finalized cases per interviewer. However, it was necessary to also validate non-respondents; which is to our knowledge, not generally done in German social surveys. Thus, supplementary control mechanisms were implemented. Apart from hard refusals, which cannot be legally re-approached in Germany due to data protection laws, an attempt was made to validate as many other dispositions as possible, such as disabled non-respondents, soft refusals etc.[6] The final disposition of a case was documented in its case folder.[7] In addition, the interviewers provided information relevant for further contact attempts. The information on the case folder was used to verify whether a contact was actually made or attempted as reported by the interviewer.

Non-respondents with literacy-related or disability-related dispositions were sent a short control questionnaire by mail (including for example age and gender and asking for a confirmation of the interviewer's contact). In order to validate some of the soft refusals, the following in-person validation[8] were put in place: As part of the refusal conversion strategy, an interviewer change was accomplished for a certain number of cases. The new interviewer was asked to try to validate the contact attempt of the previous interviewer as far as they could without compromising the main goal of obtaining an interview. For cases in which the interviewer documented that the sampled person had moved, the registry was re-contacted and requested to provide the new address. If a new address was confirmed by the registry, the previous information given by the interviewer had been validated. An interviewer then tried to interview this person at the new address. If the same 'old' address was confirmed by the registry, interviewers had to re-attempt to locate the sampled person. A random selection of cases with a final disposition code indicating that the sampled person lived in an institution were verified by checking whether there was in fact an institution located at this address. Finally, to ensure that no disposition code was used exceptionally often the frequency distribution of disposition codes was analyzed per interviewer.

6 Only one disposition was not validated: death.

7 The case folder was an important document containing the contact protocol and other relevant information about the respondent provided by the interviewer.

8 In-person validation was recommended by the Consortium.

- Review of Tape Recordings

To comply with the PIAAC *Standards and Guidelines* – each interviewer in Germany had to record two interviews on tape. The interviewers were instructed to do the tape recordings early in the field period, but only if a respondent agreed to record the interview. The tape recordings were audio monitored and evaluated by the GESIS-Team. The evaluation focused on whether the behavior of the interviewer was professional and in accordance with the standards throughout the whole interview – for example, did they read the exact wording of the questions, did they influence the routing of the questionnaire by recording answers in an incorrect way and did they use correct probing techniques. Any noticeable errors in their behavior discovered in the tape recordings were reported to the survey organization. The identified interviewers received feedback immediately and were re-trained. A positive side effect of the tape recordings was that the recorded interview is very unlikely to be a falsification.[9]

- Look at the Overall Pattern of an Interviewer's Performance

Of course, interviewers' performance was monitored throughout the whole field period. To look at the pattern of interviewers' work – as recommended by the Consortium – more than the internationally required interviewer and interview quality control checks were carried out nationally. Different sources available were used to look at the overall interviewers' performance. For this reason, for example, response rates, number of completed or unworked cases and routing checks were monitored per interviewer and feedback was given regularly to interviewers by the survey organization if the results were below or above a predefined criteria.

- Validation of 100% of the Cases for Interviewers with Suspect Cases

The control procedures foresaw that if a falsification or seriously suspicious work would be identified during the quality control checks, all worked cases by those interviewers would have to be scrutinized in detail. In Germany no falsifications could be identified so far. Thus, no validation of 100% of the cases for any interviewer was necessary.

9 A falsification is very unlikely as the information collected in the tape recordings were also compared to information from the registry. It is very unlikely that a friend interviewed instead corresponds exactly to the age and gender etc. as the sampled person.

- Additional Control Mechanisms in Germany

To cover as many areas of interviewers' work as possible, the following additional control procedures were carried out:

a) The date and time at which interviews took place were checked to see whether they were realistic. Additionally, interview length was checked with a focus on the very short interviews to ensure that these were not falsifications. Interview length can be considered as a good indicator of interviewers' performance (ESS 2011). If any outliers – especially very short interviews – were identified, the whole interview data was checked in detail (e.g. with regard to biographical consistency). As the control questionnaire also contained information about the timing estimates by the respondents, the information from respondents could be compared to the actual timing information. Moreover, interviewers were usually contacted and asked to provide more information about a specific interview.
b) Spot checks of the routing in the background questionnaire and the assessment were carried out for each interviewer to see whether they used shortcuts to reduce the time needed for an interview.
c) Quality checks on open questions were also conducted at the beginning of the field period. Responses to questions on occupations used for ISCO-coding were validated to ensure that the information which interviewers had recorded was appropriate.[10] Approximately 1,700 open answers were checked for insufficient information or duplications, as the latter might indicate a falsification. None of the answers recorded were seen as insufficient and no duplications from the same interviewer were found. Therefore, the control procedures which were not a standard but only suggested by the Consortium were also followed with regard to this aspect.
d) Interviewers had to "score" some answers given by respondents in the paper-based assessment which respondents with insufficient computer experience received. This means that interviewers had to assign codes based on the responses given by the respondent to eight quite simple questions. The value of the code corresponds either to a correct answer, an incorrect answer or no answer given by the respondent. Interviewer scoring was a crucial task as the routing in the paper assessment depended on it. At the beginning of the field period a sample of approximately 150 paper booklets was re-scored by the GESIS-Team to validate the quality of the interviewers' performance of scor-

10 In case of doubt, the national coders who would have to code open answers after the field period were consulted to ensure that the data quality was adequate.

ing. Although a few small errors were identified, none of them were systematic and in general interviewer scoring seemed to work well.

Interviewer Training for PIAAC

We have introduced quality control procedures that can be carried out during or after data collection. However, we consider it essential to already start the quality assurance process before data collection begins. In the data collection process, interviewer training is one of the first steps in ensuring high data quality (ASA 1997). For PIAAC extensive in person interviewer trainings were carried out.[11] Therefore, the next section focuses on important aspects of the trainings for PIAAC.

According to the *best practices for surveys* by the American Association of Public Opinion Research (AAPOR 2010) a careful interviewer training on interviewing techniques and the subject matter of the survey is recommended. Interviewer training aims at preparing the interviewers for the survey and at standardizing survey procedures across interviewers and (in the case of international studies such as PIAAC) across countries. In such a training the procedures of the study are explained and trained in detail to ensure that interviewers record information correctly and in the same standardized manner from each respondent. In addition, interviewer training can help to provide interviewers with effective strategies on gaining respondents' participation and on refusal prevention and/or conversion. Research has shown that interviewer training does indeed have a positive impact on response rates: Both Billiet and Loosveldt (1988) and Groves and McGonagle (2001) found that the cooperation rate is higher for trained interviewers.

PIAAC contained several innovative and challenging features (e.g. new software, new laptops, a direct competencies assessment and its partial scoring through the interviewers themselves) which were also new to interviewers highly experienced in conducting social science surveys. These aspects made a thorough training on all tasks related to PIAAC essential. Therefore, each of the 129 appointed interviewers for PIAAC in Germany had to take part in a 5-day-training before starting data collection.[12] The trainings were conducted in small

11 Interviewer trainings as extensive as in PIAAC are to our knowledge uncommon in Germany. In several important German surveys no interviewer training or trainings with an average duration of 2 days are conducted: ESS (ESS 2012), NEPS (Aust et al. 2011), SHARE (Schröder 2011), SOEP (SOEP 2011).

12 Some of the interviewers already worked for the PIAAC Field Test and had received a 5-day-training before the Field Test. Therefore, for the Main Study they only had to attend a 3-day-refresher training.

groups to maximize their efficiency. The major goal of the trainings was to illustrate and explain all materials and procedures of PIAAC and to further give the interviewers the opportunity to practice the implementation through role plays while trainers could monitor their behavior and give feedback. The interviewer trainings in Germany were held in accordance with the *Standards and Guidelines*. They consisted mainly of sessions on specific aspects of PIAAC such as a general introduction to PIAAC and the implemented software, an extensive introduction to the background questionnaire and the competencies assessment. Some sessions were adapted and extended to the national context. Other aspects of the trainings were respondent recruitment strategies and addressing respondents' concerns related to PIAAC. Moreover, the interviewers had the opportunity to share their "best practices" in gaining respondents' cooperation.

As the disposition codes were one of the central aspects of quality control and monitoring, and the PIAAC disposition codes had some specific features not shared by other national surveys, interviewers were briefed in detail about how to use the correct disposition codes for all potential cases. Some other technical and administrative issues were brought up during the trainings to refresh the interviewers' knowledge and to ensure that they were well prepared for all aspects of the PIAAC survey.

During the trainings the interviewers were informed about the measures of control that the survey organization and the GESIS-Team would carry out during the field period, such as the checks performed on the records of open answers, the audio-tapes and the other control instruments addressed above.[13] A high degree of transparency was thus achieved, and interviewers were at the same time made aware that data quality and control would be taken very seriously for each step in PIAAC. Interviewers were informed that both the survey organization and the GESIS-Team would have a close look at their performance and the quality of the collected data.

Although the primary goal of the interviewer trainings was to provide interviewers with sufficient knowledge and skills to work on PIAAC, there were several additional benefits from the trainings. According to the training evaluation feedback questionnaire, interviewers perceived the trainings as a very valuable experience and as an essential tool to best prepare them for working on PIAAC. At the end of the training period interviewers reported that they felt very confident in administering this complex and demanding study.

Another benefit we observed was that the interviewer trainings were a key factor in motivating interviewers for PIAAC in Germany. We are under the im-

13 As already mentioned this was also a requirement by the general standards described by the International Consortium.

pression that the high degree of motivation shown by PIAAC interviewers stems from the fact that interviewers and researchers had the opportunity to discuss and share experiences during the trainings. This interaction was especially valuable because interviewers rarely have contact with researchers (if at all). Interviewers felt respected and could see that their opinion and experience was valued. The 5-day trainings also reduced anonymity and were confidence-building on all sides. Interviewers reported that the training had increased their motivation and their commitment to PIAAC. Through this commitment to the project, it can be assumed that interviewers were less likely to falsify data and are more motivated to collect data of the highest quality possible.

Discussion and Conclusion

In this chapter the quality control mechanisms in PIAAC were introduced. As PIAAC is an international study, comparable data is of major importance. The Consortium for PIAAC therefore specified detailed standards to which all participating countries had to comply. The main focus of this article was on how the internationally specified *Standards and Guidelines* for quality control concerning interviewers' work were implemented in Germany. We described how these standards were adapted to the national context in Germany and which supplementary control mechanisms were added nationally.

In summary, Germany widely complied with the international standards regarding interviewer and quality control in PIAAC. A control questionnaire was sent by mail to 100% of the respondents. Data from the interview was compared to data from the registry and to data indicated on the control questionnaire. Controlling completed interviews seemed to be most important as the chance of falsification and poor data quality is the highest for completed interviews. All other dispositions that we were legally allowed to follow-up were validated through other procedures, such as tracking sampled persons that had moved. With these measures the requirement of validating 10% of each interviewer's finalized cases was achieved. It was shown that further international standards and some important recommendations were also met in Germany, such as monitoring of recorded audio tapes from interviews and interviewers' performance was analyzed on a regular basis. We also illustrated some more control procedures added nationally, e.g. timing and routing checks.

The package of control measures described above represents a further development of control procedures in Germany to ensure that data quality was the highest priority in PIAAC.

Controlling interviewers' work is important. However, this control is only carried out once data is collected. Interviewer trainings were described as a measure of ensuring data quality already prior to data collection. With the elaborate interviewer trainings for PIAAC, new paths were explored in the German survey context.

With the implementation of such a variety of different and detailed procedures, data quality assurance was brought to a new level in Germany. In total, the described quality control mechanisms as well as the interviewer trainings seemed to have a positive impact on data quality. We suspect that this might even have prevented falsifications.

Bibliography

AAPOR (American Association for Public Opinion Research): http://www.aapor.org/Best_Practices1.htm#top, revised May 2010, accessed on 21 February 2012.

ASA (American Statistical Association): How to Collect Survey Data, ASA: Section on Survey Research Methods, 1997.

Aust, Folkert, et al.: Methodenbericht NEPS Etappe 8, Befragung von Erwachsenen Haupterhebung 1. Welle 2009/2010 (35-39), Bonn, 2011.

Billiet, Jacques and Loosveldt, Geert: Improvement of the quality of responses to factual survey questions by interviewer training, Public Opinion Quarterly Volume 52 (190-211), 1988.

ESS (European Social Survey): European Social Survey: Informationen zur fünften Welle der deutschen Teilstudie (3), http://www.europeansocialsurvey.de/dokumentation/fuenfte.ueberblick.pdf, accessed on 09 July 2012.

ESS (European Social Survey): Round 6 Specification for Participating Countries, London: Centre for Comparative Social Surveys, City University London, 2011.

Groves, Robert M. and McGonagle, Katherine A.: A theory-guided Interviewer Training protocol regarding survey participation, Journal of Official Statistics Volume 17 (249-265), 2001.

Koch, Achim: Gefälschte Interviews: Ergebnisse der Interviewerkontrolle beim ALLBUS 1994, ZUMA Nachrichten 36 (89-105), 1995.

Lyberg, Lars E. and Biemer, Paul P.: Quality assurance and quality control in surveys, In: De Leeuw, Edith D., Hox, Joop J. and Dillman, Don A. (eds.): International Handbook of Survey Methodology (421-441), New York/London: Lawrence Erlbaum Associates/Taylor & Francis, 2008.

OECD (Organization for Economic Co-operation and Development): PIAAC Technical Standards and Guidelines, Paris, 2011.

OECD (Organization for Economic Co-operation and Development): PIAAC Technical Standards and Guidelines, Paris, 2008: http://www.oecd.org/officialdocumentsearch/0,3673,en_2649_201185_1_1_1_1_1,00.html?doRecentDocsSearchNoGrid=1&recentDocsYear=2008, accessed on 21 February 2012.

Schaeffer, Nora C., Dykema, Jennifer and Maynard, Douglas W.: Interviewers and Interviewing. In: Marsden, Peter V. and Wright, James D. (eds.): Handbook of Survey Research (437-470), Bingley: Emerald, 2010.

Schröder, Matis (ed.): SHARELIFE Methodology, Mannheim, 2011.

SOEP (Sozio-oekonomisches Panel): Interviewer training workshop at TNS Infratest headquarters in Munich, SOEP Newsletter 92 (9), 2011.

Detecting Interviewer's Deviant Behavior in the Slovenian National Readership Survey

Ana Slavec and Vasja Vehovar

Abstract

National Readership Surveys (NRS) are among the largest probability-based commercial surveys. Their readership estimates for national newspapers and magazines are very important for the media and advertising industry. As in other surveys, data quality in NRS could be jeopardized by interviewer fabrication of questionnaires or other types of deviations from designed guidelines and instructions. Such irregularities can be detected with several methods, most effectively with re-contact procedures. The Slovenian NRS uses telephone re-contacts but from 2007 to 2010 it was submitted to some supplementary control procedures. First, a mail re-contact survey was implemented based on 20% of respondents' samples. The survey asked respondents about the interviewer's visit as well as some control questions from the original survey. Second, a pre-questionnaire form and a control telephone survey for non-respondents were introduced to gain insight on refusals, non-contacts and other eligible but non-responding units. The results of the re-contact survey show that only a few respondents were not really interviewed, and the accordance of answers with the survey database is quite high, but for some interviewers, a higher rate of anomalies was perceived. In addition, analysis of non-respondents indicated that most of them are eligible and could be interviewed with more resources invested in fieldwork efforts to prevent interviewers from omitting cases.

Acknowledgments

We wish to thank the Slovenian Advertising Chamber, the commissioner of the National Readership Survey, who allowed the data to be used in this publication, and Valicon d.o.o., the company that administers the survey and collects data, for sharing information and assistance with using the data.

Introduction

Quality of data collected from interviewers heavily depends on interviewer behaviour and how they adhere to designed guidelines and instructions. Thus, it is very important to proactively seek to detect any falsification or other type of interviewer misconduct that could contaminate survey data. According to the guidelines of the American Statistical Association (AMSTAT 2003) survey, falsification is defined as departures from data collection protocols, unreported by the interviewer, and includes fabricating all or part of an interview, deliberately misreporting disposition codes, miscoding answers to questions (to avoid follow-up questions), interviewing non-sampled persons or otherwise intentionally misrepresenting the data collection process. Reed and Reed (1997) introduced the idea of statistical process control to survey research by charting lower and upper control limits and analysing mean values of components of the process and their variances.

The AMSTAT guidelines (2003) list three groups of methods for detecting interviewer falsification: observational, re-contact and data analysis procedures. In observational methods, another staff member observes the interaction between interviewers and respondents, which is mostly used in telephone surveys. On the other hand, re-contact methods are usually used for face-to-face surveys. First of all, re-contact questionnaires are used to determine if the unit was interviewed; however, some factual questions are also usually posed, such as household composition, mode of data collection, length of interview, topics discussed, key items, etc. (AMSTAT 2003).

The re-contacts are most effectively done face-to-face, but this mode is also the most expensive; mail re-contacting is the least expensive, but the response rates are low. One of the earliest studies using mail re-contact was done by Hauck (1969), who sent respondents a thank you letter with a return postcard asking questions about survey participation—about one half responded. The letter was also sent to a sample of non-respondents—less than a tenth returned the postcard. In both cases, people who returned postcards were not representative of the total sample. Moreover, the value of responses is questionable since almost 15% of the non-respondents who returned the postcard reported being interviewed when they had not; meanwhile, for respondents, the length of interviews did not match in almost 40% of cases, and demographic data did not match in almost 10% of cases (Hauck 1969). Thus, mixed-mode re-contact procedures are recommended as well as using phone re-contacting when possible (Hauck 1969; AMSTAT 2003).

The selection of units to be verified by observational and/or re-contact methods, typically 5 to 15% of interviews, can be random, targeted or a combination of both (AMSTAT 2003). Random selection is designed to estimate the frequency of falsification, while targeted selection aims to identify suspicious cases and is usually more effective (Bredl et al. 2012). How to optimally design re-contact samples to detect falsification is discussed by Biemer and Stokes (1989), Bushery et al. (1999) and Jianzhu et al. (2009).

The third group of procedures, data analysis methods, can identify outliers by comparing interview length, disposition codes, production rates and responses to key questionnaire items (AMSTAT 2003). For instance, Murphy et al. (2004) show how regular reviews of interview response distributions, question-level timing data and metadata can help identify cheating interviewers. Falsification can be detected using several statistical approaches, such as clustering (Bredl et al. 2013, this volume; Storfinger and Winker 2012), discriminant analysis (Bredl et al. 2013, this volume) and Benford's Law (Swanson et al. 2003; Bredl et al. 2013, this volume), which is a popular tool to detect fraudulent data not only in sample surveys, but also more generally. Furthermore, data mining techniques are applied as a falsification detection tool (e.g., Murphy et al. 2005), but literature on this topic is very scarce since analyses are often business confidential and not available for public review (Murphy et al. 2005).

According to Bredl et al. (2012), very little research has been done on the effects of data falsification on statistical analyses. Although fabrications are rare events (most often below 5%), and cheating interviewers are good at reproducing univariate distributions from real data, it was found that falsification can impact results of multivariate statistical analysis as shown by Schraepler and Wagner (2003). Thus, it is very important to monitor interviewer performance and minimize deviant behaviour.

Background

National Readership Surveys (NRS) are among the largest syndicated commercial surveys. The aim of these continuous, probability-based samples is to provide reliable readership estimates for national newspapers and magazines. Serving as a currency to buy and sell advertising space in print media, they are very important for the media and advertising industry.

The Slovenian National Readership (http://www.nrb.info/) survey started in 2002 as a computer-assisted interviewing (CATI) survey, but since 2006, data has been collected with computer assisted personal interviewing (CAPI). Every three months, a gross quarterly sample of 5,000 units with known name, home address, gender and age is drawn from the Central Population Register of the Statistical office. A two-stage probability sampling design is used with census districts as primary sampling units and individuals aged 10-75 as secondary units. The design ensures representativeness on quarterly level and uniform daily samples. With response rates around 35%, the net sample is around 1,700 per quarter and 6,500 per year. Personal interviewing is performed by 10 full-time employed interviewers and 15 students working part-time.

The interviewers of the Slovenian NRS are regularly monitored by the fieldwork coordinator. All questionnaires by newly recruited interviewers are reviewed, and afterwards, usually since the third month, depending on the results, 10 to 15% of all completed interviews are examined. In addition, every month a sample of 10 to 15% of respondents is selected to be authenticated by phone. These respondents are asked if they were visited by the interviewer, how long was the interview, did the interviewer try to manipulate answers and some socio-demographic characteristics to be matched with data from the register. Each discrepancy is thoroughly checked in accordance with ISO standards. The previous work of the interviewer is re-controlled, and in case of serious and frequent discrepancies, a replacement follows. Records are kept for all examinations, but the results are not publicly available.

As a result of deviations in readership estimates when moving from CATI to CAPI design, a tender to research methodological issues of the Slovenian NRS was made, and the Centre for Methodology and Informatics at the University of Ljubljana was selected in the process. The monitoring started in mid-2007 and ended in mid-2010. Several aspects of the survey were examined, but the principal control activity was regular monitoring of field interviewers with a control survey among respondents (and later also with non-respondents). The design of the study is presented in the following section.

Methodology

Control Survey of Respondents

From 2007 to 2010, the Slovenian NRS was supervised by researchers from the University of Ljubljana. A mixed design of mail, phone and paper surveying was developed to monitor a sample of respondents.

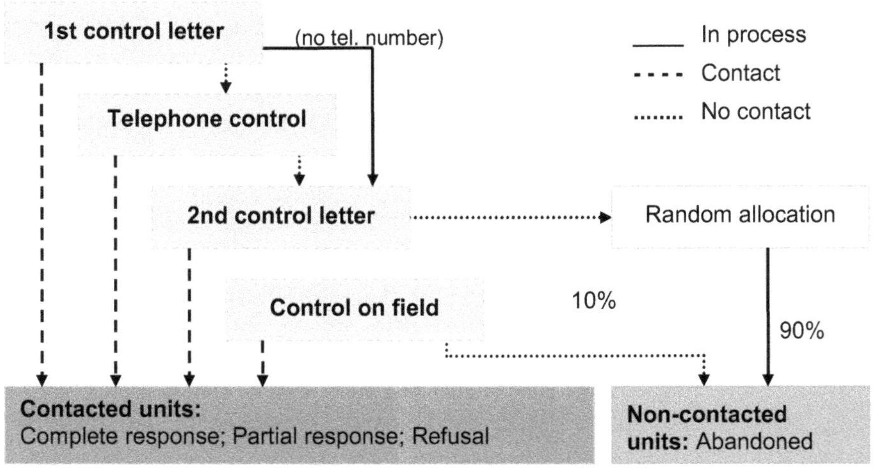

Fig. 8.1: Data collection process for control survey of respondent

Each week, a control questionnaire was mailed to 20% of respondents (about 1,200 per year) asking them about the interviewer's visit, if they were personally interviewed and how they would assess the interviewer's behaviour. To verify survey answers, some questions from the survey were also included: how often they use the Internet, to which of the eight major Slovenian newspapers is their household subscribed and what is their gender and age.

The letter with the control questionnaire was sent one to two weeks after the fieldwork interview, except for the start of the monitoring in the second quarter of 2007 when the letters were sent all at once at the end of the quarter—for some units, this would be up to three months after the fieldwork. It could be said that in the first year (2007), the control process was in pilot phase. After some slight design and questionnaire improvements in the beginning of 2008, stabilization was achieved in the following quarters. Another peculiarity in the design was in 2009—due to the decreased funding this year, the sample size was lowered to 10% of respondents. In 2010, it was back to 20%.

About half of the selected respondents answered within two weeks while others were called by phone (if number was listed), and those not contacted after five trials were sent another letter. Among the respondents that did not respond to the second letter, we randomly selected 10% to be visited by a field interviewer (Figure 8.1). In total, about 90% of the control sample cooperated; in the first two years, the rate was at 91%, while in the last year, it fell by three percentage points to 88%.

Control Survey of Non-respondents

About two thirds of all eligible units in the NRS are not interviewed due to non-response (e.g., in 2008, there were 14,242 non-interviewed units out of the 20,568 sampled). This includes refusals, non-contacts and other eligible units. It is well known that non-responding units can substantially differ from respondents and induce biased estimates. To research these differences, a pre-questionnaire form was introduced in mid-2008. Interviewers were requested to fill out the type of dwelling and ask for household subscriptions to eight major Slovenian newspapers. The type of dwelling was obtained for all visited units while the question about newspaper subscription (which could be posed also to another household member) was responded by only 60% of all eligible non-responding units.

In addition, a telephone control survey of non-respondents was implemented in 2008. Every month, a sample of 5% of units was selected to be controlled, which summed to 683 units in the whole year. For 492 units (72%), a contact number was found listed in the telephone register. However, only 456 (67%) of them remained after removing non-working and otherwise non-eligible telephone numbers. Then, 406 units (59%) were successfully contacted and 356 households (52%) cooperated, i.e., answered control questions.

Non-respondents were surveyed for two main purposes. First, to establish the eligibility of units (eligible, non-eligible or unknown eligibility), which is important to accurately compute response rates, and second, to collect data on non-respondents (socio-demographic characteristics and newspaper subscription rates) to use to compute non-response bias.

A standardized procedure to determine eligibility was developed and applied on the sample. If the telephone number listed was shown to be incorrect, the unit was labelled with "unknown eligibility." If the target person was successfully contacted (cooperating or not), it was indicated as eligible. When the target person could not be contacted, determining eligibility was more complicated. Those that did not pick up phone kept the unknown eligibility status. When another person from the phone responded but did not answer the control survey, we used information gained in the conversation to establish eligibility. For instance, if we were informed that the target person is permanently ill, not capable to participate or in an institution (e.g., hospital, prison), it was marked as non-eligible. If none of this was the case, but discovered the target person does exist although non-reachable, it was marked as eligible. The determined eligibility of units using this procedure was compared with the eligibility status reported by interviewers. In addition, for units that responded to the control survey, the match of the disposition code was also controlled, and they were asked if they received the survey pre-notification letter.

Results

Control Survey of Respondents

In total, 90% of the control sample cooperated in the control survey. The results are shown in Table 8.1. The amount of deviations oscillates around 3%; about 1% affirmed they were not interviewed, while others are interviewed by phone (completely or partially) or do not remember being interviewed.

In the first year of the project, almost 4% of the surveyed respondents claimed to have not been interviewed; almost 5% were interviewed by phone (at least partially) while more than 1% did not remember the interview. In total, almost 10% of the sample was problematic. The rates were particularly high in the second quarter when the monitoring started, but this irregularity can be explained with the recalling problems—respondents received the letter up to three months after being interviewed. Another problem was poor wording of some control questions, which was resolved in the following years.

After the first year, the amount of irregularities decreased, in the end oscillating around 3%. In 2008, the rate of non-interviewed fell from almost 4% to 2%, and approximately 1% in 2009 and 2010. Even bigger was the decrease in the percentage of units interviewed completely or partially by phone. This is also partially due to changes in the wording of the question about being interviewed by phone. Since 2008, the "no" category has been split into "No, the phone was used only to set the appointment," and "No, I never spoke with the interviewer by phone." This was done to prevent categorizing appointment interviews partially done by phone, which was obviously the case in 2007.

Table 8.1: Irregularities found in the Slovenian Readership Survey

Year and Quarter		Not interviewed (%)	Telephone (%)	Telephone partially (%)	Does not remember (%)	Total deviations (%)	*Responded to control*	*Included in control sample*
2007	q2	6.0	3.5	3.5	3.2	16.2	*286*	*329*
	q3	2.2	1.9	2.4	0.2	6.7	*412*	*433*
	q4	2.4	1.7	1.9	0.7	6.7	*417*	*447*
	Total	3.7	2.2	2.5	1.2	9.6	*1115*	*1209*
2008	q1	1.3	0.6	0.3	0.6	2.9	*308*	*327*
	q2	0.4	0.4	0.0	1.5	2.2	*272*	*315*
	q3	3.1	0.0	1.0	2.1	6.2	*291*	*323*
	q4	2.5	0.7	0.7	1.8	5.6	*284*	*312*
	Total	1.8	0.4	0.5	1.5	4.2	*1155*	*1277*
2009	q1	2.8	0.7	2.1	0.7	6.2	*146*	*164*
	q2	0.7	0.0	0.0	0.0	0.7	*150*	*165*
	q3	0.7	1.5	0.0	2.2	4.4	*136*	*160*
	q4	1.8	0.0	0.0	0.0	1.8	*165*	*191*
	Total	1.3	0.5	0.7	0.7	3.2	*597*	*680*
2010	q1	0.7	1.1	0.4	0.4	2.8	*285*	*341*

Of course, the decreased number of deviations after 2007 is connected with interviewers being aware of the controls. The release of the first control results affected the behaviour of both survey staff and interviewers who strived to improve performance. There were also some fluctuations among quarters. In the summer months (third quarter), there are usually more irregularities, which is probably because of the vacation times of interviewers as well as of respondents. The deviation rate is also increased in the last quarter as the end of the interviewing year is approaching and delayed appointments with respondents cannot be made. On the other hand, the deviations in the first quarters were usually due to modifications (improvements) in the design of the control survey.

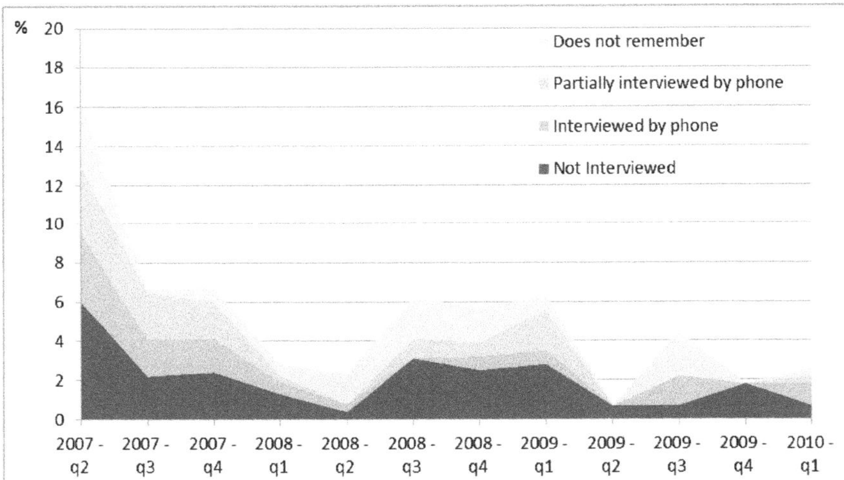

Fig. 8.2: Irregularities found in the Slovenian Readership Survey

Analysis by Interviewer

The number of interviewers in the Slovenian NRS is quite low, which could be problematic because of intraclass correlations caused by interviewer bias and enhanced by the design decision to assign each interviewer to a specific geographic region (to diminish costs). This causes difficulties in comparing interviewer performance and their deviation. However, indicative results are presented in Table 8.2.

There are eleven interviewer "positions," which cover definite geographic areas. The first ten are occupied by full-time employed interviewers, while the eleventh (two biggest towns) is covered by part-time working interviewers (students). The table presents how many interviews were performed, the average response rate, irregularities detected with re-interviewing and the average evaluation rate they received from respondents who answered the control survey.

Two interviewers obtained a significantly higher response rate than others (A1, A4), while one interviewer deviated in the opposite direction (A7). There were also differences in regards to deviations from survey procedures (non-interviewing, interviews by phone, unknown status). Three interviewers in 2007 (A1, A4, and A9), two in 2008 (A1 and A6) and one in 2009 (A8) had significantly more irregularities than others. The rate of deviations shows a decreasing trend. In regards to respondent satisfaction with the interviewer, there were not many differences among interviewers—it has been consistently high at 4.8 out of 5 throughout the years.

Table 8.2: Comparison of interviewers

Interviewer	Number of interviews			Response rate (%)			Total deviations (abs. freq.)			Average interviewer evaluation (on scale 1-5)		
	2007	2008	2009	2007	2008	2009	2007	2008	2009	2007	2008	2009
A1	1434	1611	1656	47.0	41.2	40.4	9	7	2	4.7	4.8	4.7
A2	/	309	2605	/	36.9	28.2	/	1	0	/	4.5	4.7
A3	1755	1862	2156	40.9	36.3	32.2	1	1	2	4.9	4.9	4.9
A4	1572	1657	1811	43.3	39.8	41.4	11	4	2	4.9	4.8	4.8
A5	/	1392	2095	/	39.3	33.4	/	2	0	/	4.7	4.7
A6	1699	2085	1820	40.7	29.5	29.3	2	6	0	4.8	4.7	4.9
A7	2105	2172	2274	22.7	20.1	13.6	3	4	1	4.8	4.8	4.8
A8	/	1775	2274	/	30.5	27.6	/	3	7	/	4.8	4.8
A9	1378	1775	2627	36.7	30.5	25.7	9	2	0	4.8	4.8	4.8
A10	/	1665	2528	/	29.3	28.5	/	5	1	/	4.6	4.7
Other (A2, A5, A8, A10)	5,492	2,196	59	37.8	25.7	22.0	15	3	0	15	4.8	4.8
Students	2745	1601	1590	26.6	27.4	36.6	66	11	4	66	4.9	4.8
Total	6,405	6,316	6,925	35.5	31.1	29.8	116	49	19	116	4.8	4.8

Matching of Control Variables

First, accordance between the NRS database and the Central Population Register was checked for gender and age (Table 8.3). In the beginning (2007), accordance was only 90%, but it improved to 95% in the end. Other units are mostly unknown (incomplete data in NRS database, particularly in 2007); less than 1% of units do not match.

Table 8.3: Accordance between NRS and Central Population Register for gender and age

	2007	2008	2009
Matching on gender	89.8%	93.9%	94.9%
Discrepancy on gender	0.4%	0.7%	0.5%
Missing data	9.7%	5.4%	4.6%
Matching on age (one year of difference tolerated)	89.0%	93.9%	94.5%
Discrepancy on age (two or more years difference)	1.1%	0.5%	0.8%
Missing data	9.9%	5.5%	4.7%
Total units	100%	100%	100%
	6,409	6,314	6,924

In 2009, matching was also compared by interviewers, and it was found that one interviewer (A9 in Table 8.2) had produced significantly more discrepancies in gender and/or age than others. On the other hand, for two interviewers (A4 and A7 in Table 8.2), almost no discrepancies in control variables were found. It may also be noted that one of these two (A7) is the interviewer that attained a significantly lower response rate than others.

Table 8.4: Accordance between NRS and Control Survey for Internet use and newspaper subscriptions

	2008	2009
Matching on Internet use	69.0%	51.0%
Discrepancy on Internet use	13.6%	32.9%
Missing data	17.4%	12.8%
Matching on newspaper subscriptions	64.1%	73.6%
Discrepancy on news subscriptions (on at least one newspaper)	3.1%	13.6%
Missing data	32.7%	12.8%
Total units	100%	100%
	675	678

Note: In 2008, the data was collected for 20% of respondents in the second half of the year–the question was included in the NRS survey in August 2008. In 2009, data was collected throughout the whole year, but only for 10% of the respondents.

Furthermore, matching for Internet use and newspaper subscription was examined based on data from the control survey of respondents (Table 8.4). Here, accordance with NRS data is not as high as for gender and age from the register. In 2009, only half of the respondents reported the same rate of Internet use in the control survey as in the NRS, while a third indicated a different category. The match for newspaper subscriptions was 74%; meanwhile, for 14% of cases, responses differed for at least one newspaper. For the remaining 13% of cases, there is no data in one or both of the sources. In 2008, there were fewer discrepancies both for Internet use (14%) and especially for newspaper subscriptions (3%), but this may be connected with the higher rate of missing data in this year since the question had only recently been included in the original NRS questionnaire. In addition, sometimes interviewers forgot to fill out the pre-questionnaire during the first months it was introduced. Another difference between 2008 and 2009 is that in 2008, data was not collected throughout the year, but only for five months.

However, not all the discrepancies in Internet use and newspaper subscription rates are necessarily due to interviewer deviations. To a certain extent, they result from other sources of measurement error, in particular, recall problems. In fact, the respondents were re-contacted from a week to up to two months after the survey, and they might not give the same answer as the first time. Moreover, Internet use habits could also change in the time between the interview and the re-interview. This could also be applied to newspaper subscriptions, but to a smaller extent since this characteristic is less likely to change in such a short period of time. It should also be noted that discrepancy in Internet usage is partially due to the difference in question wording; the question in the original survey had eight response categories, which were re-coded to match the five categories used in the control survey.

Non-respondents

A control survey was also performed on non-respondents, but of course, the procedure was different than for respondents. The focus was on determining their eligibility and comparing it with the information provided by the interviewer. In fact, it has been shown that most of the non-interviewed units (81%) are actually eligible. Not surprisingly, the percentage is highest among refusals (86%) and non-contacts (83%). However, what is interesting is that it is also quite high among units where the interviewer reported that nobody was home at any visit (unknown eligibility), where we were able to successfully contact 78% of the cases and mark the unit as eligible. For units that changed residence, the percentage of eligible units is a bit lower, 67%, while a third of them are really non-eligible. Moreover, four of the five non-eligible units were also found to be eligible, but the number is too small to generalize. On the other hand, 1.4% of controlled non-responding units were found to be non-eligible, and for 18%, it was not possible to determine their eligibility.

For 356 units, which are a little more than half of the units selected for the control survey, answers to control questions were also attained. Only 20% of them remembered receiving the NRS pre-notification letter while 41% did not remember it, and 39% stated they did not receive it. Most of them were among the non-contacted units. Next, the responses to control questions show that about 13% of the controlled households disagreed with the interviewer's final outcome status. In most cases, it was about refusals: 22% of them claim they did not refuse to participate in the survey. Moreover, among units non-contacted in the original survey, 8% said they were not out of reach during the two-week fieldwork period when the interviewer was assigned to their district.

In addition, target persons (or other household member if the target person could not be contacted) also responded to a control question about newspaper subscriptions. By adding these answers to respondents' answers, non-response bias was computed. A comparison of subscription rates with and without respondents is shown on Figure 8.3, where subscription rates for respondents are given on x-axis while estimated subscription rates for total sample (respondents and non-respondents) are given on y-axis; each symbol represents one newspaper. For instance, the newspaper with more than 14% of subscribers according to NRS estimates has less than 12% of subscribers if we make an estimate merging the NRS data with answers of non-respondents. The same is also true for other newspapers. Evidently, by omitting non-respondents, the NRS underestimates subscription rates.

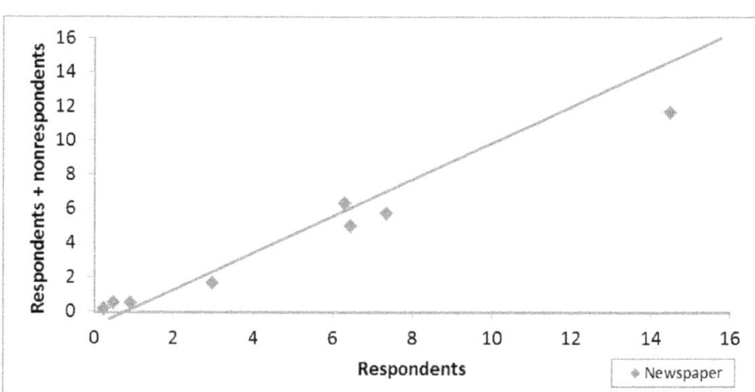

Fig. 8.3: Comparison of subscription rates of respondents only and respondents merged with non-respondents

Conclusions and Discussion

To a certain extent, data quality in the Slovenian National Readership Survey could be compromised due to error produced by interviewers who fabricated all or parts of the answers to the questionnaire or otherwise departed from data collection procedures. In fact, results of the control survey—to which 90% of the re-contact sample cooperated—show that approximately 3% of respondents were not really interviewed or were interviewed by phone. Analysis by interviewers indicates some of them had significantly more anomalies than others. In addition, there are some discrepancies in key variables—some of the units have reported different frequencies of Internet use and newspaper subscriptions than in the original survey. On the other hand, accordance with register data of gender and age is very high.

Another source of error arises from reporting disposition codes and omitting eligible units from the survey. More than every fifth refusal denies that he did not want to be interviewed, and one in twelve non-contacts disavowed being hard to reach in the fieldwork period. Non-respondents are a problem as they usually differ from respondents, and if this difference is big and the response rate is very low, estimates can be substantially biased. In the case of national readership surveys, subscriptions are probably overestimated. This is a topic that will be addressed in a separate paper.

Despite the rate of irregularities not being very high, more resources could be invested in fieldwork efforts to prevent interviewers from deviating. Here, basic techniques to detect survey falsification are presented based on the case of the Slovenian NRS. While nothing is said about how to correct for this error, what are the effective actions to take when there is evidence of fabricated survey data, and which procedures should be used to document and disclose detected falsifications. In future research, more attention should be placed on motivating and building commitment among personnel. Survey practitioners also need guidelines on how to correct falsified data.

Bibliography

AMSTAT: Interviewer Falsification in Survey Research: Current Best Methods for Prevention, Detection and Repair of Its Effects, 2003. http://www.amstat.org/sections/srms/falsification.pdf

Biemer, Paul and Stokes, Lynne: Optimal Design of Quality Control Samples to Detect Interviewer Cheating, Journal of Official Statistics, Vol. 5 (1), pp. 23-39, 1989.

Bredl, Sebastian. Winker, P., Kötschau, Kerstin: A Statistical Approach to Detect Interviewer Falsification of Survey Data. Survey Methodology, forthcoming, 2012.

Bredl, Sebastian, Storfinger Nina and Natalja Menold.: A Literature Review of Methods to Detect Fabricated Survey Data. In: Winker, Peter, Natalja Menold and Rolf Porst (Eds.): Interviewers' Deviations in Surveys – Impact, Reasons, Detection and Prevention. Peter Lang, Frankfurt a. M., 2013, pp. 3-24.

Bushery, John M., Reichert, Jennifer W., Albright, Keith A. and Rossiter, John C.: Using date and time stamps to detect interviewer falsification. Proceedings of the American Statistical Association, Section on Survey Research Methods, 1999.

Hauck, Mathew: Is survey postcard verification effective? Public Opinion Quarterly 33:117-20, 1969.

Jianzhu, Li., Brick, Michael, Tran, Bac and Singer, Phyllis: Using Statistical Models for Sample Design of a Reinterviewer Program. Joint Statistical Meeting, Section on Survey Research Methods, 2009.

Murphy, Joseph., Baxter, Rodney K., Eyerman, Joseph D., Cunningham, David and Kennet, Joel: A System for Detecting Interviewer Falsification. Paper presented at the 59th annual conference of the American Association for Public Opinion Research, 2004. http://www.rti.org/pubs/paper-detectingintrvfalsifi.pdf

Murphy, Jospeh, Eyerman,, Joseph D., McCue, Colleen, Hottinger, Christy and Kennet, Joel: Interviewer falsification detection using data mining. Proceedings of Statistics Canada Symposium, 2005. http://www.statcan.gc.ca/pub/11-522-x/2005001/4199022-eng.pdf.

Reed, Stephanie J., & Reed John H.: The Use of Statistical Quality Control Charts in Monitoring Interviewers. Proceedings of the American Statistical Association, Section on Survey Research Methods, 1997.

Storfinger, Nina and Winker, Peter: Robustness of Clustering Methods for Identification of Potential Falsifications in Survey Data, 2012.

Swanson, David, Cho, Moon J., and Eltinge, John L.: Detecting Possibly Fraudulent or Error-Prone Survey Data Using Benford's Law. Proceedings of the American Statistical Association, Section on Survey Research Methods, 2003.

Vehovar, Vasja., Slavec, Ana., Makovec, Meta and Hren, Mateja: Nadzor nad Nacionalno raziskavo branosti, Poročilo za leto 2007, Ljubljana: Fakulteta za družbene vede, 2008.

Vehovar, Vasja., Slavec, Ana., Makovec, Meta and Hren, Mateja: Nadzor nad Nacionalno raziskavo branosti, Poročilo za leto 2008, Ljubljana, Fakulteta za družbene vede, 2009.

Vehovar, Vasja., Slavec, Ana., Javoršek, Katja and Hren, Mateja. Nadzor nad Nacionalno raziskavo branosti, Poročilo za leto 2009, Ljubljana, Fakulteta za družbene vede, 2010a.

Vehovar, Vasja, Slavec, Ana and Javoršek, Katja. (2010b): Nadzor nad Nacionalno raziskavo branosti. Poročilo za leto 2010, Ljubljana, Fakulteta za družbene vede, 2010b.

III Discourses on Interviewers Behavior and Deviations in Survey Data

Interviewer Behavior and the Quality of Social Network Data[1]

Josef Brüderl, Bernadette Huyer-May, Claudia Schmiedeberg

Abstract

Interviewer effects are a typical – although often neglected – phenomenon of social network data collected in personal interviews. We analyze the ego-centered network data provided by the German Family Panel and find large interviewer effects which cannot be explained by interviewer or respondent characteristics. These interviewer effects are caused to a large degree by two groups of interviewers, i.e. those who elicit less network persons than the average ("fraudulent" interviewers) and those who generate particularly large networks ("diligent" interviewers). We suggest a method to identify these groups of interviewers.

Introduction

In recent years, social network analysis has gained popularity (for an overview see e.g. Wassermann and Faust 2009). The majority of social network studies have relied on data gathered by standardized personal interviews (Marsden 1990, Matzat and Snijders 2010). But caution is advised as social network data are vulnerable to "noise" (Fischer 2009) as well as to interviewer and design effects. These effects may result from deviating interviewer behavior. In contrast to faked interviews, which most of the other chapters of this volume (e.g. Bredl et al. 2013, Menold et al. 2013) deal with, in our case the interviewers' deviations are limited to a small part of the interview whereas other parts of the data should not be affected. Nevertheless, according to AAPOR (2003) this can be seen as a (partial) falsification of interviews. In the research tradition on falsifications generated by interviewers a large body of literature on "curbstoned" interviews exists (e.g. Crespi 1945, Evans 1961, Biemer/Stokes 1989, Schräpler/Wagner 2005), whereas other types of deviating interviewer behavior have been in the focus only rarely.

In this contribution we investigate interviewer effects found in ego-centered network data from the second wave of the German Family Panel. Our results reveal strong interviewer effects in the network data which cannot be explained by given interviewer characteristics such as sex or age. In addition to quantifying the interviewer effect we identify deviating interviewers by using the jack-

1 This paper uses data from the German Family Panel (pairfam), coordinated by Josef Brüderl, Johannes Huinink, Bernhard Nauck, and Sabine Walper. pairfam is funded as a long-term project by the German Research Foundation (DFG).
For helpful suggestions we are grateful to Frauke Kreuter.

knife. Using this method, we find that the interviewer effect is a consequence of both interviewers eliciting a below-average number of alters and those generating particularly large networks.

The chapter proceeds as follows: In the next section we discuss the reasons for interviewer effects in social network data. The section "Data and Methodological Issues" gives details on the pairfam data and the name generators, our empirical strategy and the variables used. The following section presents our results and the last section concludes.

Reasons for Interviewer Effects in Social Network Data

Deviant interviewer behavior resulting in interviewer effects is a well-known phenomenon. Questions have been raised regarding the role of interviewer and respondent characteristics on the size of interviewer effects as well as on aspects of type and form of questions and interviewer training. Groves and Magilavy (1986) report that interviewer effects are more likely to occur for attitudinal questions (in contrast to factual ones), in particular in case of open-ended responses (see also Mangione et al. 1992), emotional topics, difficult items and when "the acceptable interviewing procedure" (Groves and Magilavy 1986: 260) is not specified. A large interviewer effect can be expected also for questions asking the respondent to list a (not predetermined) number of entities. Schnell and Kreuter (2005) assume a combination of all these factors to affect the probability of interviewer effects.

Interviewer effects in social network data have been shown in the literature before (van Tilburg 1998, Straits 2000, Marsden 2003, Wolf 2004). In addition, social network data can be seen as particularly prone to interviewer effects due to the typically difficult data collection process: The most common method for the measurement of ego-centered social networks is the use of name generators (Laumann 1966, Burt 1984): First, the respondents are asked to list individuals with whom they are in specific types of relationships (alters), before in a next step a number of questions follows which gather information about the generated alters. This procedure of eliciting names is burdensome for both respondents and interviewers (Marsden 2003, McCarty et al. 2007). As a consequence the potential lack of informant accuracy has been discussed in the literature, i.e. the phenomenon that respondents are not able to recall the alters who meet the relationship criterion (Marsden 1990, Marin 2004). Interviewer behavior, such as probing, might influence the respondent's efforts (Marsden 2005: 17). On the other hand, given the fact that also well trained interviewers are not throughout able to abide by the rules defined for the interview, e.g. reading questions exactly as printed (Groves and Magilavy 1986), differences in the interviewers' ability might influence the results (Billiet and Loosveldt 1988, Van Tilburg 1998, Straits 2000, Marsden 2003).

Another obvious mechanism that might produce interviewer effects was put forward by Matschinger et al. (2005): rational interviewer behavior. The collection of network data is time consuming as each alter produced by a name generator leads to a number of follow-up questions. Thus, interview duration can be reduced by producing fewer alters. As interviewers are typically paid per interview instead of getting an hourly wage, rational interviewers might try to shorten the interview by entering fewer names than the respondent would name. In the extreme they will skip the entire network module. If we have a mixture of diligent and rational interviewers much of the variation in the number of alters will be due to differing interviewer behavior.

Data and Methodological Issues

Our analysis is based on the network data from the second wave of the German Family Panel (release 2.0), which were collected in 2009/2010. pairfam is a representative longitudinal study for researching partnership and family dynamics. It started in 2008 with a random sample of more than 12,000 individuals of the three birth cohorts 1971-73, 1981-83 and 1991-93. A detailed description of the pairfam study can be found in Huinink et al. (2011). Data are collected via a one hour CAPI interview with a CASI part for sensitive topics. Questions about the respondent's social network were included in the CAPI in wave 2 for the first time; they were placed quite at the end of the interview.

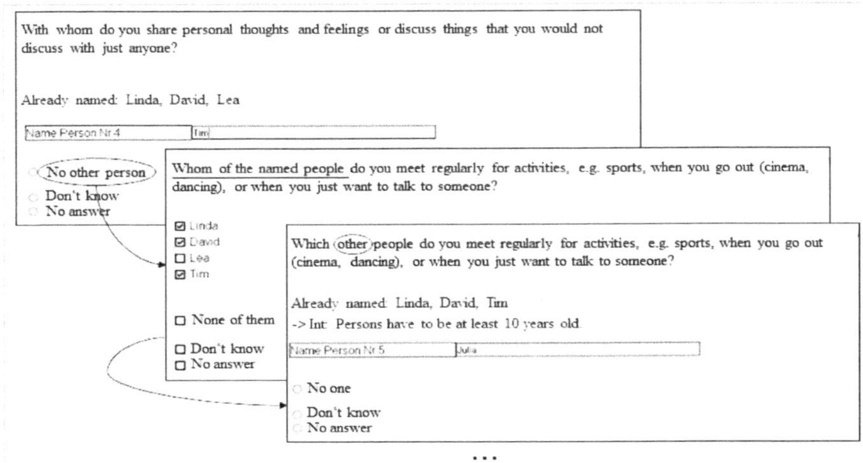

Figure 9.1: The name generator in pairfam

The network module in pairfam consists of four name generators, followed by 12 descriptor questions, of which the first two are asked for all alters named and the others only for a maximum number of 8 randomly chosen alters. The four name generators are the following:

1. With whom do you share personal thoughts and feelings or discuss things that you would not discuss with just anyone?
2. Which persons do you meet regularly for activities, for example sports, or when you go out (cinema, dancing), or when you just want to talk to someone?
3. Who helps you whenever you need information or concrete advice in practical matters?
4. With whom do you occasionally have quarrels or conflicts?

The respondents are asked to write down the names of the alters on a sheet of paper and then tell them to the interviewer, who types them in after each generator.[2] A maximum of 30 contacts can be entered per generator, so that a total of 120 network contacts can be gathered. However, only 30 different names of alters are allowed. When a name is given in more than one generator, instead of entering the name a second time the interviewer just ticks the respective box in the list of names already generated (see Figure 9.1).

2 In contrast to other parts of the CAPI, probing is not prescribed in the network module.

In the following we will analyze the number of alters generated by the four generators (network size), which is the quantity used in most substantive studies. Alternatively, one could investigate the number of network contacts generated. However, generating alters is more time consuming than generating contacts so that we expect more pronounced interviewer effects in the first case.

As an adequate method to disentangle interviewer and respondent effects a hierarchical regression model is used (Hox 1994). When the explanatory variables at the individual level are denoted as $X_1,\ldots,X_p$, and those at the interviewer level $Z_1,\ldots, Z_q$ the model can be written as follows (Snijders and Bosker 1999: 51):

$$Y_{ij} = \gamma_{00} + \gamma_{10}\, x_{1ij} + \ldots + \gamma_{p0}\, x_{pij} + \gamma_{01}\, z_{1j} + \ldots + \gamma_{0q}\, z_{qj} + U_{0j} + R_{ij}$$

To assess the amount of interviewer effects we focus on the intraclass correlation coefficient (ICC) indicating the share of the variance that can be attributed to the interviewers (Kish 1962). A high ICC implies that answers from respondents are more similar when they are interviewed by the same interviewer (West and Olson 2010). The ICC (ρ), is defined as

$$\rho = \sigma^2_{U_0} / (\sigma^2_{U_0} + \sigma^2_R),$$

where σ^2_U is the variance due to the interviewers and σ^2_R is the respondent-level variance.

We include a number of control variables in the analysis. On the respondent level these are age, sex, partnership status, existence of children in the household and employment status. Unfortunately, we do not have a lot of information on the interviewers. As control variables we include age, sex and a dummy indicating if the interviewer has conducted interviews also in the first wave of pairfam. In order to take into account interaction effects between respondents' and interviewers' sex, we consider interaction terms of these variable as well. Besides, we include the sequence number of each interview (within an interviewer) to account for possible learning effects.

In a next step, we jackknife our estimation (for a general description see for example Efron 1982 or Shao and Wu 1989); that means, we analyze how much the ICC is reduced when all interviews of a particular interviewer are excluded. By this we identify which interviewers are most responsible for the large ICC.

Results

Tables 9.1 and 9.2 present some descriptive figures on the pairfam data. Average network size (Table 9.1, row 4) is smaller than we would have expected given the broad types of everyday behavior covered by the four name generators. As anticipated, we see that younger respondents generate slightly larger networks, but especially for the youngest cohort we would have expected much larger networks.

Table 9.1: Respondent Characteristics

	Cohort 1	Cohort 2	Cohort 3
Years of Birth	1991-1993	1981-1983	1971-1973
% Partnership	33.5%	69.9%	82.4%
% Females	49.4%	51.0%	56.0%
Network size	4.77 (3.32)	4.21 (3.16)	3.82 (2.93)
Number of observations	3,555	2,621	2,893

For network size mean values and standard deviation are reported.

Regarding the interviewer characteristics (Table 9.2) the wide range as well as the mean value of the interviewers' age is notable. Also the number of interviews conducted differs strongly, spanning from 1 to 137 (6% of the interviewers completed more than 70 interviews).

Table 9.2: Interviewer Characteristics

	Interviewers
% Females	42%
% Experience from wave 1	83%
Age	58.90 (10.06)
Number of interviews per interviewer	26.59 (22.42)
Number of observations	341

For interviewers' age and the number of interviews per interviewer mean values and standard deviation are reported.

Table 9.3 shows the results of the multilevel regressions on network size. The first model does not contain any control variables, model 2 includes variables on the respondent level, and model 3 includes both respondent and interviewer characteristics as well as interaction effects.

A main finding is the remarkably high ICC in all models, which implies that interviewer behavior strongly influences the number of alters generated. Our ICCs are around 0.4. The literature reports for roughly comparable network data from surveys somewhat smaller ICCs: van Tilburg (1998) reports an ICC of 0.21, Marsden (2003) for the General Social Survey (GSS) 1998 0.15, and Wolf (2004) for his own survey 0.28. Obviously interviewer effects in pairfam are particularly large.

Intraclass correlation can be reduced only to a small degree by including control variables. The ICC even increases when including respondent characteristics only, which points to the fact that the interviewer effect is not caused by the uneven distribution of respondents to the interviewers. The ICC also does not decrease if we include the sample points to control for regional differences (results not shown). Thus, the interviewer effects are not due to regional effects.

We do not go into detail regarding the control variables, but three interesting effects should be noted: First, our results indicate an inverse U-shaped relationship between the number of interviews an interviewer conducted and the number of network persons he or she elicited on average. That means interviewers with both a low and a high number of interviews tend to generate smaller networks than those with a medium number of interviews (the "peak" is obtained with 53 interviews). Second, the interviewers' age has a highly significant effect, indicating that older interviewers generate smaller networks.

Third, we find a positive, significant learning effect (see the effect of "sequence number"). This means that interviewers elicit a larger number of alters as they get accustomed to the handling of the name generators in the course of the field. This contradicts the main finding of Matschinger et al. (2005) who report that the effect of sequence number in their study is negative. They argue that this is due to "rational" interviewer behavior who learn to shorten interviews by reducing the number of alters generated. With pairfam data we are able to reproduce their result, but when controlling for respondents' age the effect of sequence number becomes positive, as reported in Table 9.3. The spurious result found in the bivariate analysis is due to the fact that younger respondents – who have on average larger networks – were in the field earlier because they are easier to contact.

Table 9.3: Multi-Level Regressions on the Network Size

	Model 1	Model 2	Model 3
Characteristics interview			
Sequence number (within interviewer)		0.0042*	0.0068*
		(2.50)	(2.57)
Characteristics respondent			
Activity status (Ref.: School)			
Vocational training		-0.485***	-0.481***
		(-5.28)	(-5.24)
College, university		-0.020	-0.008
		(-0.16)	(-0.06)
Full-time employed		-0.240*	-0.229*
		(-2.36)	(-2.24)
Part-time employed		-0.478***	-0.469***
		(-3.71)	(-3.63)
Housewife / househusband		-0.649***	-0.639***
		(-4.69)	(-4.61)
Unemployed		-0.940***	-0.927***
		(-7.02)	(-6.92)
Age		-0.029***	-0.028***
		(-4.99)	(-4.86)
Partner		0.089	0.090
		(1.48)	(1.49)
Female		0.570***	0.642***
		(10.29)	(9.15)
Children		-0.445***	-0.447***
		(-5.45)	(-5.47)
Characteristics interviewer			
Number of interviews			0.032*
			(2.48)
Number of interviews squared			-0.0003*
			(-2.09)
Female interviewer			-0.058
			(-0.25)
Age interviewer			-0.047***
			(-4.17)
Experience from wave 1			-0.086
			(-0.92)
Female interviewer x female respondent			-0.175
			(-1.71)
Number of respondents	9069	9069	9069
Number of interviewers	341	341	341
ICC	0.404	0.416	0.397

Notes: The effects reported are unstandardized. The values in parentheses contain the z-values of the estimation. *** p<0.001, **p<0.01 and *p<0.5

Next we try to identify "deviant" interviewers. First, we want to identify those interviewers who show rational behavior and shorten or even skip the network generators. We call these the "fraudulent" interviewers. Second, there are those interviewers who have a particularly high motivation. They try to elicit every network contact (probably by probing intensively). We call them the "diligent" interviewers. Finally, there is the majority of "normal" interviewers who do their job more or less reasonable. They work strictly according to rule, i.e. they do not probe and are satisfied if a few alters are named.

To identify these groups we suggest using the jackknife.[3] We estimate model 1 from Table 9.3 by dropping all interviews of one interviewer and record the reduction in the ICC. This step is repeated for all interviewers. To concentrate on the most deviant interviewers, we set the threshold at the 95%-percentile. Thus, we identify the 5% interviewers who affect the ICC most (5%-sample). Table 9.4 gives statistics for four groups of interviewers: (1) the full sample, (2) 95%-sample (those who are not in the 5%-sample), (3) interviewers from the 5%-sample with below average network size and (4) interviewers from the 5%-sample with above average network size. Group (2) comprises the "normal" interviewers, group (3) consists of the "fraudulent" interviewers, and group (4) includes the "diligent" interviewers.

Concerning the descriptives we see that the diligent interviewers are on average younger and conduct fewer interviews. The fraudulent interviewers on the other side conduct much more interviews. Both fraudulent and diligent interviewers are mostly male.

Table 9.4 reveals large differences regarding the mean network size: The fraudulent interviewers produce an average network size of 1.6, an remarkble small number. By looking on the networks produced by some of these interviewers, we identified several of them, who mainly produced networks of zero size. For instance, interviewer 1324 (18 interviews) showed the following sequence: 0, 1, 2, 1, 1, 0, …, 0. Obviously, this interviewer learned after the first five interviews that it is much more convenient to skip the network module altogether.

The diligent interviewers, in contrast, generate on average 12 alters. This, in our opinion, seems to be a realistic network size given the young age of our respondents. This means that the normal interviewers with an average network size of 4.3 alters are still much below the "real" network size. Only diligent interviewers, probably by intensive probing, elicit even the last alters.

---

3 For a description of the jackknife and its applications see Efron (1982) and Shao and Wu (1989).

Further insight can be gained by looking on the number of alters generated by the four generators separately (Table 9.4, rows 5-8). Most alters are generated by the first generator ("personal feelings"). The diligent interviewers generate also many alters on the second generator ("meeting"). This is what one would expect especially with young people going out often. Not surprisingly the third ("helping") and fourth ("conflict") generator do not generate many additional alters. Contrary to this, the fraudulent interviewers generated hardly ever new alters in the second, third or fourth generator. We interpret this again as rational behavior in the sense that interviewers skip the module after the first generator.

Table 9.4: Descriptives and Network Size for Four Groups of Interviewers

	Full sample	**"Normal" int.** (95%-sample)	**"Fraudulent" int.** (5%-sample below average)	**"Diligent" int.** (5%-sample above average)
Number of interviews per interviewer	26.60 (22.42)	26.30 (21.71)	47.14 (43.14)	18.50 (20.83)
Age	58.85 (10.06)	58.87 (10.03)	63.00 (5.13)	53.33 (14.60)
% Females	43.10%	44.20%	14.29%	16.67%
Network size	4.30 (3.22)	4.31 (3.01)	1.62 (1.49)	12.02 (5.56)
1st Generator	2.56 (1.30)	2.56 (0.90)	1.08 (0.66)	6.35 (1.90)
2nd Generator	1.16 (1.03)	1.16 (0.96)	0.28 (0.31)	4.18 (1.39)
3rd Generator	0.43 (0.40)	0.43 (0.35)	0.21 (0.18)	1.09 (0.62)
4th Generator	0.15 (.17)	0.15 (0.15)	0.05 (0.06)	0.44 (0.30)
Number of interviewers	341	328	7	6
Number of interviews	9069	8628	330	111

Means and standard deviations reported.

In addition, analyzing the generators separately enables us to compare our data with other studies, since most other studies used only one generator, which is comparable to our first generator. Our first generator is similar to the one introduced by Burt (1984) in the GSS 1985 and has been used in several studies subsequently (Campbell and Barrett 1991). In our data the network size produced by the Burt generator is on average 2.6 (full sample). This is very similar to the figures found in prior studies: In the GSS 1985 the mean network size turned out to be 2.9. Bien et al. (1991) report four German surveys using the Burt generator with mean numbers of alters ranging from 1.8 to 2.6.

However, as argued above we believe that the mean network size of 2.6 in pairfam is a gross underestimate. We suppose the "real" network size to be more than double (6.34 produced by the diligent interviewers). We even think our finding can be generalized: Because most large scale network surveys are troubled with interviewer effects, network sizes in most studies might be underestimated due to interviewer behavior.

Conclusion

We analyzed the social network data provided by the German Family Panel and found significant interviewer effects that cannot be explained by the given characteristics of interviewers or respondents. An ICC of about 0.4 says that almost half of the variation in network size is not "real" but is due to more or less careful interviewer behavior. Further, the most diligent interviewers produced networks that are three times as large as the average. Thus, not only a large part of variation is artificial but also the mean is downward biased.

As mechanism behind this we suppose rational interviewer behavior. Interviewers are paid per interview and thus have an incentive to keep the interview short. Compared to "normal" interviewers who needed on average 6 minutes to produce 4.3 alters "fraudulent" interviewers saved 4 minutes by largely skipping the network module. In contrast, the "diligent" interviewers - producing on average 12 alters - needed 15 minutes for the network module. Collecting the full networks of respondents is obviously very time-consuming and usually this time is not paid for. Thus, it comes as no surprise that some interviewers chose the "rational" way and shortened the interview by reducing the network module to the first generator or even skipping it. In this regard the effects we found are rather similar to the cases of faked interviews presented e.g. by Menold and Opper (2013, this volume) or Gwartney (2013, this volume).

Probably this is not only a phenomenon of the pairfam network data. Other surveys report average network sizes that are similarly small. We estimate that the "real" network sizes are double or even triple in size. Further, in other network data also large ICCs of around 0.2 have been reported. Thus, we suppose that all network data where interviewers have the incentive to shorten the interview are threatened by interviewer behavior. Others (e.g. Marsden 2003, Wolf 2004) have made similar arguments. Hence, caution is advised when using network data collected by personal interviews, in particular regarding mean network size. The question is, however, if correlational analyses are affected as well. If interviewers reduce networks more or less randomly, one might contend that regression models are not affected. For instance, one could argue that the results reported in Table 9.3 seem reasonable according to what is known from the network literature so that the large intraclass correlation in our data does not seem to bias correlational results. However, it is obvious that this is a heroic argument.

Finally, what can be done to collect more valid network data? Behavioral theory identifies rational interviewer behavior as the underlying problem. So either one should use diligent interviewers only, or one should apply incentives that provoke the desired behavior (see Wolf 2004 for similar arguments).

Using diligent interviewers only usually would mean that the network researcher does the survey by herself (or at least with hand-picked interviewers). Wolf (2004) provides evidence that self-collected network data show much lower intraclass correlation. However, this is not a real option in most cases. Then, a more intense interviewer training perhaps could provide more motivated interviewers. Interviewers should be made aware of the aims, scope and limitations of the network data and train the handling of the name generators. If necessary there should be even repeated interviewer training.

Another way would be setting the incentives right. Crespi (1945) has identified several aspects of survey design which might affect the interviewers' propensity to deviate. According to his suggestions, a lengthy network module at the end of an extensive interview may represent the typical setting for interviewer deviations. Secondly, paying interviewers by the hour would be one option that, however, is not available with most commercial survey agencies (at least in Germany).

Strengthening interviewer control would be another option. Wolf (2004) argues that network data collected by carefully supervised interviewers show larger networks and the ICC is smaller. He provides some evidence that due to this mechanism (and because of the hourly pay scheme) better network data are collected by telephone interviews. pairfam has collected network data again in wave 4 with the same instruments and field procedures. A field monitoring procedure was established whereby interviewers who generated networks with only zero or one alter in several interviews were contacted and asked for explanation during the field time. It remains to be seen if this control mechanism leads to improved data quality.

Bibliography

AAPOR (American Association for Public Opinion Research) (2003): Interviewer Falsification in Survey Research: Current Best Methods for Prevention, Detection and Repair of Its Effects. Retrieved June 18, 2012, from http://www.amstat.org/sections/srms/falsification.pdf.

Biemer, Paul P. and S. Lynne Stokes (1989): Optimal design of quality control samples to detect interviewer cheating. Journal of Official Statistics 5: 23-39.

Bien, Walter, Jürgen Marbach and Franz Neyer (1991). Using egocentered networks in survey research. A methodological preview on an application of social network analysis in the area of family research. Social Networks 13(1): 75-90.

Billiet, Jacques and Geert Loosveldt (1988): Improvement of the Quality of Responses to Factual Survey Questions by Interviewer Training. Public Opinion Quarterly 52(2):190-211.

Bredl, Sebastian, Nina Storfinger and Natalja Menold (2013): A Literature Review of Methods to Detect Fabricated Survey Data. In: Winker, Peter, Natalja Menold and Rolf Porst (Eds.): Interviewers' Deviations in Surveys – Impact, Reasons, Detection and Prevention. Frankfurt/Main: Peter Lang. pp. 3-24.

Burt, Ronald S. (1984): Network Items and the General Social Survey. Social Networks 6(4): 293-339.

Campbell, Karen E. and Barrett A. Lee (1991): Name generators in surveys of personal networks. Social Networks 13(3): 203-221.

Crespi, Leo P. (1945): The Cheater Problem in Polling. Public Opinion Quarterly 9(4): 431-445.

Efron, Bradley (1982): The jackknife, the bootstrap and other resampling plans. Philadelphia: Society for Industrial and Applied Mathematics.

Evans, F. B. (1961). On Interviewer Cheating. Public Opinion Quarterly 25: 126-127.

Fischer, Claude S. (2009): The 2004 GSS Finding of Shrunken Social Networks: An Artifact? American Sociological Review 74(4):657–669.

Groves, Robert M. and Lou L. Magilavy (1986): Measuring and Explaining Interviewer Effects in Centralized Telephone Surveys. Public Opinion Quarterly 50(2): 251-266.

Gwartney, Patricia A. (2013): Mitschief versus Mistakes: Motivating Interviewers to Not Deviate. In: Winker, Peter, Natalja Menold and Rolf Porst (Eds.): Interviewers' Deviations in Surveys – Impact, Reasons, Detection and Prevention. Frankfurt/Main: Peter Lang. pp. 195-215.

Hox, J. J. (1994): Hierarchical Regression Models for Interviewer and Respondent Effects. Sociological Methods & Research 22(3): 300-318.

Huinink, Johannes, Josef Brüderl, Bernhard Nauck, Sabine Walper, Laura Castiglioni and Michael Feldhaus (2011): Panel Analysis of Intimate Relationships and Family Dynamics (pairfam): Conceptual framework and design. Zeitschrift für Familienforschung 23(1): 77-101.

in varying the wording of personal network name generators. Social Networks 22(2), 123–140.

Kish, Leslie (1962): Studies of Interviewer Variance for Attitudinal Variables. Journal of the American Statistical Association 57(297): 92-115.

Laumann, Edward O. (1966): Prestige and association in an urban community: an analysis of an urban stratification system. Indianapolis: Bobbs-Merrill.

Mangione, Thomas W., Floyd Fowler. and Thomas A. Louis Floyd (1992): Question Characteristics and Interviewer Effects. Journal of Official Statistics 8(3): 293–307.

Marin, Alexandra (2004): Are respondents more likely to list alters with certain characteristics? Implications for name generator data. Social Networks 26(4): 289-307.

Marsden, Peter V. (1990): Network Data and Measurement. Annual Review of Sociology 16: 435-463.

Marsden, Peter V. (2003): Interviewer effects in measuring network size using a single name generator. Social Networks 25(1): 1-16.

Marsden, Peter V. (2005): Recent Developments in Network Measurement, pp 8-30. *In:* Carrington, Peter, John Scott and Stanley Wasserman (Eds): Models and Methods in Social Network. Cambridge: Cambridge University Press.

Matschinger, Herbert, Sebastian Bernert and Matthias C. Angermeyer (2005): An Analysis of Interviewer Effects on Screening Questions in a Computer Assisted Personal Mental Health Interview. Journal of Official Statistics 21(4): 657-674.

Matzat, Uwe and Chris Snijders (2010): Does the online collection of ego-centered network data reduce data quality? An experimental comparison. Social Networks 32(2): 105-111.

McCarty Christopher, Peter D. Killworth and James Renell (2007): Impact of methods for reducing respondent burden on personal network structural measures. Social Networks 29(2): 300-315.

Menold, Natalja and Marie Kristin Opper (2013): Interviewers' Experience and "Quality" of Faked Data. In: Winker, Peter, Natalja Menold and Rolf Porst (Eds.): Interviewers' Deviations in Surveys – Impact, Reasons, Detection and Prevention. Frankfurt/Main: Peter Lang, pp. 25-47.

Schnell, Rainer and Frauke Kreuter (2005): Separating Interviewer and Sampling-Point Effects. Journal of Official Statistics 21(3): 389-410.

Schräpler, Jörg-Peter and Gert G. Wagner (2005). Characteristics and impact of faked interviews in surveys – An analysis of genuine fakes in the raw data of SOEP. Allgemeines Statistisches Archiv 89: 7-20.

Shao, Jun and Wu, C.F.J: (1989): A General Theory for Jacknife Variance Estimation. Annals of Statistics 17(3): 1176-1197.

Snijders, Tom and Bosker, Roel (1999): Multilevel Analysis. An introduction to basic and advanced multilevel modeling. London: Sage Publications.

Straits, Bruce C., (2000): Ego's important discussants or significant people: an experiment

Van Tilburg, Theo (1998): Interviewer Effects in the Measurement of Personal Network Size. A Nonexperimental Study. Sociological Methods and Research 26(3): 300-328.

Wassermann, Stanley and Katherine Faust (2009): Social Network Analysis: Methods and Applications, Cambridge: Cambridge University Press.

West, Brady T. and Kristin Olson (2010): How much of interviewer variance is really nonresponse error variance? Public Opinion Quarterly 74(5): 1004-1026.

Wolf, Christof (2004): Egozentrierte Netzwerke. Erhebungsverfahren und Datenqualität. In Andreas Diekmann (Ed.): Methoden der Sozialforschung. Sonderheft Kölner Zeitschrift für Soziologie 44: 244-273.

Interviewer Experience and "Quality" of Falsified Data

Natalja Menold and Marie Kristin Opper

Abstract

When dealing with detecting falsifications in survey data it is important to consider the characteristics of interviewers. There is some evidence that the "quality" of falsified data is particularly influenced by an interviewer's experience. We assumed that experienced interviewers are better than inexperienced interviewers at reproducing real data. In order to test this assumption, we compared real data with falsified data which was artificially generated by persons with either a lot or with relatively little prior experience as interviewers. Some indicators are characterized as formal indicators which focus on response sets, while other indicators, referred to as "content-related", take into account the responses to survey questions. The results show significant differences between experienced and inexperienced interviewers: inexperienced interviewers differ more strongly from real respondents in terms of content-related indicators, while experienced interviewers produce greater differences from real data in formal indicators. The results are discussed in terms of their usability for detection methods.

Introduction

Interviewer characteristics have been discussed as factors which might affect the results of falsifications (see Bredl et al., 2013, this volume). In this regard, interviewer gender, age, education and length of service have been tested for any correlations with falsifying behavior (Biemer and Stokes 1989; Hippler 1979; Hood and Bushery 1997; Koch 1995; Schraepler and Wagner 2003; Schreiner et al. 1988; Wetzel 2003). In particular, the length of service as an interviewer has been found to be crucial: interviewers who are likely to be dishonest also tend to be rather inexperienced, with only short work periods as interviewers. For example, Schreiner et al. (1988), referring to the results of the Interviewer Falsification Study of the U.S. Bureau of Census, reported that the average length of service as an interviewer was significantly lower for falsifiers (1.72 years) than for other interviewers (6.22 years). In addition, in the study conducted by Schreiner et al. (1988) interviewers with long and short lengths of service differed in terms of the kind of falsifications they produced: newly recruited interviewers showed a higher rate of faking of the entire questionnaire, while professionally experienced interviewers produced falsifications of only parts of questionnaires (partial falsifications). Hood and Bushery (1997) supposed that experienced interviewers do not falsify less often, but rather they use more sophisticated falsification strategies, resulting in less frequent detections through controls.

This article addresses the question of how interviewers' professional experience affects the results of falsifications. Interviewer experience can be reflected in different indicators, which are used for detection purposes, for example filter questions or extreme or middle responding (presented by Bredl at al. 2012 and Menold et al.,2013,this volume; see also Schraepler and Wagner 2003). The impact of interviewer experience on the discriminatory power of an indicator has not yet been addressed by previous research, in spite of the potentially serious consequences related to the effectiveness of detection methods. Menold et al. (2013, this volume) distinguish between two groups of indicators. The first group - called "formal indicators" - refers to response behavior, meaning the ability of falsifiers to reproduce the response patterns of real respondents (i.e. concerning item non-response or extreme responding). The second group of indicators - called "content-related" indicators - concerns the content of questions, meaning the ability of falsifiers to reproduce the responses of real respondents. The current study analyses how the data produced by interviewers with different levels of professional experience differ from real data in terms of both formal and content-related indicators.

In terms of formal indicators differences between real and falsified data are explained by a varied effort on the part of falsifiers (Menold et al. 2011; Menold et al., 2013, this volume). In general, falsifiers increase their effort to produce a "good" falsification by working more carefully than real respondents would, since falsifiers believe that this kind of behavior helps to avoid detection. This behavior is associated with a low level of satisficing (Krosnick and Alwin 1987). Satisficing refers to less careful and rather superficial information processing. However, in the case of several types of questions, for example semi-open or filter questions, falsifiers should demonstrate more satisficing behavior than real respondents since responses involving the "other, please specify" category and providing extensive information in filter questions might require a high level of cognitive effort. This high level can be easily reduced by providing a legitimate response. In contrast, it would be less burdensome for a falsifier to avoid certain kinds of response patterns while administrating multi-item sets, for example by avoiding extreme responses (cf. Menold et al., 2013, this volume).

There is some evidence that experienced interviewers may differ from inexperienced interviewers with regard to satisficing behavior. This was found using cognitive interviews within the IFIS[1] project, in which falsifiers are asked about their falsification methods (Guellner and Porst 2012). In particular, only inexperienced interviewers reported difficulties in providing information about re-

1 Funded by the DFG (German Research Foundation within SPP 1292; WI 2024/2-1 and ME 3538/1-1).

spondents' fathers, since falsifiers had to consider which occupations existed in the distant past (Guellner and Porst 2012). Questions about the occupations of respondents' fathers could be skipped in the questionnaire using a filter question. By using results from cognitive interviews, it can be assumed that experienced interviewers have a higher degree of awareness in terms of socially desirable responses by real respondents. Thus, in the cognitive interview a professional interviewer expressed that in particular items concerning attitudes toward Jews had "a difficult emotional component" and therefore the interviewer assumed real respondents would answer such questions in a socially desirable way (Guellner and Porst 2012: 17).

Differences in content-related indicators between real respondents and falsifiers can be explained by less knowledge on the part of falsifiers about real respondents' opinions and behaviors. Falsifiers draw on certain general stereotypes when falsifying an interview (Hippler 1979, Schnell 1991). On the other hand, experienced interviewers are more likely to possess accurate knowledge about respondents' opinions and behaviors, and in this way they use stereotypes less frequently than inexperienced interviewers.

As we have assumed that experienced interviewers possess more knowledge about the response behavior of respondents and that they stereotype less often when answering survey questions, we expect them to produce fewer differences from real data than inexperienced interviewers in terms of both formal and content-related indicators.

Procedure

In the study presented here we used data collected within the project IFIS "Identification of Falsifications in Survey Data". This paper is based on a bachelor thesis by Marie Kristin Opper (Opper 2011), and her work was supervised by Natalja Menold. In the study, a procedure by Reuband (1990) and Schnell (1991) was applied (Menold et al. 2011; Menold and Kemper 2011): a subsample of n = 105 interviews from an existing dataset (German General Social Survey, ALLBUS 2008) was drawn. From the ALLBUS data, only interviews which had been verified by ALLBUS control procedures as being valid were used. In addition to the real ALLBUS interviews, false data (n = 105 falsifications) were produced by 16 study participants. The study participants were instructed on the procedure and purpose of the study. They were also instructed that they were to imagine being real interviewers who are not able to conduct an interview with the intended person and are now trying to generate falsified data. In doing so, the study participants were instructed to reproduce the responses of real participants as closely as possible. They used descriptions of real respondents including several details (place of residence, age, sex, confession, number of children, party preference and education) while producing falsifications. After falsified data were generated, these were then compared with the real ALLBUS data.

A part of the study participants was comprised of experienced interviewers who were employed at the Telephone Laboratory of GESIS. They had already conducted from 60 to over 1000 interviews. The other study participants consisted of students, as in previous studies by Reuband (1990) and Schnell (1991). These participants did not have any professional experience as interviewers. We refer to the first group as "professionals" and to the second group as "students". Since two students were also found to be experienced in conducting face-to-face interviews (they had conducted about 800 interviews), they were added to the group of "professional" interviewers. This resulted in nine study participants in the "professionals" group and in seven in the "students" group. The average age of the "professionals" was roughly 41 years (SD = 9), while the "students" had a mean age of approx. 23 years (SD = 1). Six of the 16 participants were males (five professionals and one student).

A Multivariate Analysis of Variance (MANOVA) was used in the study to compare data between the three groups: real ALLBUS respondents (n = 105 interviews), "professionals" (with n = 57 falsifications) and "students" (with n = 48 falsifications). Prior to the analysis we tested the MANOVA assumptions. We were not able to confirm a normal distribution of dependent variables using the Kolmogorov-Smirnov-Test within each group. However, MANOVA was found to be robust concerning violations of multivariate normal distribution if the sample size is relatively large (n > 40) and the number of variables is relatively small (k <10) (Bortz 1999). The assumption of homogeneous variances within the compared groups was supported by the F_{max} test (Field 2009) for all dependent variables.

Operationalization of Indicators

Within the MANOVA we used six formal and three content-related indicators as dependent variables. Table 10.1 provides information regarding the operationalization of these indicators. We used extreme responding, item non-response, filter questions, rounding, tendency to disagree, and social desirability as formal indicators. The first four indicators were also used and described in detail by Menold et al. (2013, this volume). The indicator "tendency to disagree" is the counterpart of acquiescence, which describes the tendency to agree, in other words to answer "yes" regardless of the content of the item (see Bortz and Doering, 2002, for an overview). Acquiescence is particularly pronounced in cases of dichotomous questions (e.g. Krosnick and Presser 2010). Thus we used such questions to construct the indicator. To build the indicator "social desirability" we used items related to right wing extremism and anti-Semitism. We considered strong negative responses to the statements used as indicating socially desirable behavior, for example "I am proud to be a German", "Under certain circumstances a dictatorship is the best form of government" or "Jews have simply something special and peculiar, and therefore do not really suit us".

Like Menold et al. (2013, this volume), we expected study participants to respond to survey questions with a higher level of effort (less satisficing) than real respondents. This expectation applies to all formal indicators, except for filter questions. Thus we expect to obtain less extreme responding, lower item nonresponse, less rounding, lower acquiescence and less socially desirable responding in the falsified data than in the real data. In the case of filter questions, we expect study participants to reduce their effort (like Menold et al., 2013, this volume), particularly if they are inexperienced "students". Reduced effort should be visible through more frequent skipping of subsequent questions by using a filter question (see Table 10.1). However, for all indicators we expect to obtain higher differences between the "students" and real respondents, while differences between "professionals" and real respondents should be lower.

As content-related indicators we used a number of attitude and behavior variables. Since we assume that falsifiers use stereotypes, we can expect that their data differ from real data, in particular if the falsifiers are inexperienced. The "political participation" indicator used in the study (Table 10.1) has also been used by Menold et al. (2013, this volume) and refers to a multiple choice question about the past political participation of respondents. The indicator "women's labor force participation" consists of six items, for example "A child will surely suffer if his/her mother works" or "It is even better for a child when his/her mother has a job and when she does not only focus on the household". The items are formulated in a way that allows respondents to express either a liberal or a conservative attitude toward employment of women. "Job division in household" – our third content-related indicator - was constructed using eight items, and here respondents are asked whether men or women usually do such varied household tasks as preparing food, cleaning the house or doing minor home repairs. Stereotyping is seen if typical women's jobs are reported to be done by the female partners.

Table 10.1: Operationalization of dependent variables (indicators)

Indicators		Operationalization
Formal Indicators	Extreme	The usage of extreme categories (e.g."1" and "5" within a five-point rating scale) is counted for 118 scalar multi-item questions.
	Item Non-Response	The usage of item non-response categories is calculated over all items in the questionnaire, except the questions used for the indicator "filter".
	Tendency to Disagree	The frequency of choosing the category "no" is counted for four dichotomous questions.
	Filter Questions	The frequency of choosing one of the categories that allows skipping a part of questions about the father's education and occupation is counted.
	Social Desirability	The frequency of choosing the categories "strongly disagree" and "rather disagree" is counted for ten items. Both categories stand for rejection of extreme and racist formulated statements.
	Rounding	How often rounded values are given in six numerical open questions is calculated.
Content-Related Indicators	Political Participation	How many of the twelve items concerning the political participation are chosen is calculated.
	Women's Labor Force Participation	In a first step, two of the six items are recoded to measure in the same direction: "A mother should not go working". Then how often the categories "strongly disagree" and "rather disagree" are ticked is calculated. Both categories stand for a less conservative attitude towards women's employment.
	Job Division in Household	The usage of categories that stand for a stereotypical division of work in the household is calculated in four items. It is counted how often the categories "always done by the woman" and "mostly done by the woman" are ticked by the items "doing laundry", "doing the dishes" and "cleaning the house" and how often the categories "always done by the man" and "mostly done by the man" is used by the item "doing minor home repairs".

Results

In this section, results regarding the differences between real data and data produced by the "professionals" and the "students" are presented. According to Roy's test statistic ($F (2, 207) = 2.207$; $p < .001$, $\eta 2 = 0.37$) the MANOVA model indicates significant differences between the three groups that were compared. The univariate effects are also significant for all indicators (see F-values in Table 10.2), showing that there are significant mean differences between at least two treatment groups.

Table 10.2: Differences between real and falsified data (univariate test statistics)

	Indicators	Professionals *M (SD)*	Students *M (SD)*	Real Data *M (SD)*	*F*	η^2
Formal Indicators	Extreme	29.46 (9.32)	29.58 (10.16)	37.39 (11.95)	13.72*	0.12
	Item Non-Response	15.82 (14.71)	11.63 (8.19)	9.02 (11.00)	6.36*	0.06
	Tendency to Disagree	2.58 (0.78)	2.80 (0.58)	3.18 (0.72)	14.57*	0.12
	Filter Questions	0.37 (0.49)	0.33 (0.48)	0.19 (0.39)	3.63*	0.03
	Social Desirability	5.14 (3.09)	6.48 (2.19)	6.05 (2.16)	4.25*	0.04
	Rounding	1.11 (0.92)	1.31 (0.97)	1.83 (1.04)	11.06*	0.10
Content-Related Indicators	Political Participation	3.37 (1.92)	2.60 (1.23)	3.88 (2.20)	7.14*	0.07
	Women's Labor Force Participation	3.07 (2.38)	2.94 (2.33)	4.24 (1.85)	8.86*	0.08
	Job Division in Household	2.35 (1.53)	2.69 (1.60)	1,98 (1.45)	3.83*	0.04

Note. * $p < 0.05$

In order to find out which mean differences between the three treatment groups are significant, post-hoc paired comparisons with Bonferroni correction were conducted (see results in Table 10.3). In terms of most formal indicators we found that differences between "professionals" and real respondents are higher than those between the "students" and real respondents. Thus, data from "professionals" significantly differ from the real data concerning all formal indicators except for social desirability. As compared to the real data, "professionals" produced less extreme responses, more item non-response, they used filter questions more often in order to skip subsequent questions and they demonstrated less rounding. In contrast, "students" differed from real respondents only in that they showed less extreme responses and less rounding. In the case of "social desirability", there was a significant difference between "professionals" and "students". Here "students" provided the most socially desirable responses, while "professionals" had the lowest value and seemed to avoid demonstrating social desirability.

Table 10. 3: Significant Mean Differences between the Treatment Groups (Post-Hoc-Tests)

Indicators		Significant Mean Differences between the Treatment Groups		
		"Professionals" vs. Real Respondents	"Students" vs. Real Respondents	"Professionals" vs. "Students"
Formal Indicators	Extreme	-7.93***	-7.81***	-0.12
	Item Non-Response	6.81*	2.61	4.20
	Tendency to Disagree	-0.60***	-0.39	-0.21
	Filter Questions	0.18*	0.15	0.04
	Social Desirability	-0.91	0.43	-1.34*
	Rounding	-0.72***	-0.52**	-0.21
Content-Related Indicators	Political Participation	-0.51	-1.28**	0.76
	Women's Labor Force Participation	-1.17*	-1.30*	0.13
	Job Division in HH	0.37	0.71*	-0.34

Note. *** $p < .001$; **$p < .01$; *$p < .05$

These results do not support our initial expectation that – in contrast to the "students" – "professionals" generate data which is closer to that of real respondents when formal indicators are considered. In contrast, in terms of response behavior data from "students" are quite similar to data from real respondents. However, when considering both "professionals" and "students", it can be seen that deviations from the real data were similar in the case of most formal indicators, with both groups showing less satisficing (see Table 10.3). Thus, both groups produced less extreme responding, less rounding and a lower tendency to disagree. The result concerning the filter indicator supported our a-priori assumption regarding a lower level of effort (and more satisficing) for both groups of falsifiers as well. Hence, in the case of social desirability a higher level of effort (less satisficing) was seen only in the "professionals". In terms of item non-response we cannot support the assumption of lower satisficing on the part of falsifiers from either the "students" or the "professionals" groups.

When considering formal indicators, we found that "professionals" differed more strongly from real respondents than the "students" did. The opposite results were found with respect to content-related indicators. Thus, a significant mean difference between "professionals" and real respondents existed only in the case of "women's labor force participation". In contrast, "students" differed from real respondents in all three content-related indicators. It can be seen from Table 10.2 that professionals were better at predicting past political participation, while students underestimated this behavior more strongly. Concerning the indicator "Job division in household" students assigned typical women's jobs to female partners more often than real respondents did. Finally, "students" were more conservative in terms of women's labor force participation when compared to real respondents.

Taking all indicators together, we evaluate which of these are particularly affected by interviewer experience. We conclude that there is an effect if "professionals" significantly differ from "students", and/or a group ("students" vs. "professionals") significantly differs from real respondents, while the other group does not. This applies to nearly all formal indicators, except for "extreme responding" and "rounding" and to content-related indicators, except for "women's labor force participation" (Table 10.3).

In summary, the results show an effect of interviewer experience on the majority of indicators. We also found that data from the "students" deviated from the real data to a greater extent, but only if content-related indicators were considered. Thus, experienced interviewers were rather able to "forecast" the responses of real respondents to certain survey questions. In contrast, when considering formal indicators experienced interviewers were less able to reproduce real response behavior when compared with the inexperienced students.

Conclusions

The aim of the analysis presented in this article was to show the effects of interviewers' professional experience on the results of data falsification. In our analysis we compared real survey data with falsified data produced partly by professional, experienced interviewers and partly by inexperienced students.

The results show an effect of interviewer experience on a majority of the indicators tested in the study. Only extreme responding, rounding and attitudes toward women's labor force participation were found to differ between real and falsified data, regardless of interviewers' experience. For both groups of indicators tested – formal and content related indicators – we expected experienced interviewers to generate data that was closer to that of real respondents than inexperienced students could generate. However, as shown by our results experienced interviewers were only able to correctly reproduce real responses concerning certain behaviors and opinions, but they were unable to correctly demonstrate real response behavior. In contrast, inexperienced students differed less from the real respondents with respect to their response behavior, while at the same time they were less able to reproduce real data concerning opinions and behaviors. Next, we found that experienced interviewers were more likely to behave in line with assumptions of the satisficing model than inexperienced students – namely to show higher satisficing for filter questions and lower satisficing for most remaining formal indicators. Thus, it is important to consider interviewer experience when using the satisficing model to explain differences between real and falsified data.

If we try to evaluate the sensitivity of indicators with regard to interviewer experience, we can expect that experienced interviewers will stay undetected if content-related indicators are used. In contrast, students will remain undetected if certain formal indicators (item non-response, tendency to disagree and filter questions) are used. However, when considering real survey situations in which experienced interviewers are involved, we can conclude that formal indicators should be powerful in terms of detecting falsifications in the context of real surveys. Next, it is important to combine various formal indicators within a multivariate approach to compensate for the sensitivity of several indicators related to interviewer experience. Finally, detection strategies which use different indicators in different experience groups of interviewers or which allow for different types of falsifications would be useful. Hence, more research is needed to develop these approaches.

However, we should be critical in mentioning that the results of our study have some limitations. We only used a small number of interviewers in the groups of falsifiers. Small group sizes may also be responsible for less significant differences between the "professionals" and "students" since the Bonferroni correction requires large groups in order to achieve a high level of test power. Moreover, professional interviewers and students differed in terms of their age, and this may provide an alternative explanation for the differences between these two groups. Thus, more studies are needed which compare responses and response patterns at the interviewer level using large groups of experienced and inexperienced interviewers, as this allows for control of different explanatory variables.

Bibliography

Biemer P. P., Stokes S. L.: The Optimal Design of Quality Control Samples to Detect Interviewer Cheating, Journal of Official Statistics, 5(1), 23-39, 1989.

Bortz, J.: Statistik für Sozialwissenschaftler (Fifth Edition), Springer, Berlin, 1999.

Bortz, J., Doering, N.: Forschungsmethoden und Evaluation für Human- und Sozialwissenschaftler, Springer, Heidelberg, 2002.

Bredl, Sebastian, Storfinger Nina and Natalja Menold: A Literature Review of Methods to Detect Fabricated Survey Data. In: Winker, Peter, Natalja Menold and Rolf Porst (Eds.): Interviewers' Deviations in Surveys – Impact, Reasons, Detection and Prevention. Peter Lang, Frankfurt a. M., 2013, pp. 3-24.

Bredl, S., Winker, P., Koetschau, K.: A Statistical Approach to Detect Interviewer Falsification of Survey, Data Survey Methodology, 2012.

Field, A.: Discovering Statistics Using SPSS (Third Edition), Sage, Los Angeles, 2009.

Guellner, G., Porst, R.: Identifikation von Fälschungen in Surveys, Bericht über den kognitiven Test im Rahmen der Vorstudie 2, GESIS Working Papers, 16, 2012.

Hippler, H.-J.: Untersuchung zur "Qualität" absichtlich gefälschter Interviews, ZUMA-Arbeitspapier, 1-30, Mannheim, 1979.

Hood C., Bushery J. M.: Getting More Bang from the Reinterview Buck: Identifying "At Risk" Interviewers, Proceedings of the American Statistical Association (Survey Research Methods Section), 820-24, 1997.

Koch, A.: Gefälschte Interviews: Ergebnisse der Interviewerkontrolle beim ALLBUS 1994, ZUMA-Nachrichten, 36, 89-105, 1995.

Krosnick, J. A., Alwin, D.: An evaluation of cognitive theory of response-order effects in survey measurement, Public Opinion Quarterly, 51, 201-219, 1987.

Krosnick, J. A., Presser, S.: Questionnaire design, Wright, J. D., & Marsden, P. V. (Eds.), Handbook of Survey Research (Second Edition), Emerald Group, West Yorkshire, England, 2010.

Menold N., Kemper, C.: Response style characteristics as indicators for detection of falsifications, 4th Conference of the European Survey Research Association (ESRA), Lausanne, July 20th, 2011.

Menold, N., Storfinger, N., Winker, P.: Development of a method for ex-post identification of falsifications in survey data, NTTS, Conference Paper 4, Session 9, 2011.

Opper, M. K.: Aufdeckung von Fälschungen in Umfragen: Effekt der Interviewerfahrung auf die Genauigkeiten der Vorhersagen und die Strategie beim Erzeugen gefälschter Daten, not published BA thesis manuscript, 2011.

Reuband, K.-H.: Interviews, die keine sind. "Erfolge" und "Mißerfolge" beim Fälschen von Interviews, Kölner Zeitschrift für Soziologie und Sozialpsychologie, 42(4), 706-733, 1990.

Schnell, R.: Der Einfluss gefälschter Interviews auf Survey Ergebnisse, Zeitschrift für Soziologie, 20(1), 25-35, 1991.

Schraepler J., Wagner G.: Identification, Characteristics and Impact of Faked Interviews in Surveys - An analysis by means of genuine fakes in the raw data of SOEP, IZA Discussion Paper Series, 969, 2003.

Schreiner I., et al.: Interviewer Falsification in Census Bureau Surveys, Proceedings of the American Statistical Association (Survey Research Methods Section), 491-496, 1988.

Tourangeau, R. et al.: Sources of error in a survey of sexual behavior, Journal of Official Statistics, 13, 341-365, 1997.

Wetzel, A.: Assessing the Effect of Different Instrument Modes on Reinterview Results from the Consumer Expenditure Quarterly Interview Survey, Proceedings of the American Statistical Association (Survey Research Methods Section), 4508-4513, 2003.

Interviewers' Question: Rewording Not Always a Bad Thing

Marieke Haan, Yfke Ongena and Mike Huiskes

Abstract

Although correct question reading is a fundamental assumption of standardized interviewing, in surveys, interviewers will not always read all questions exactly as worded. In this study the deviations in question reading by interviewers were analyzed. In addition, we studied if these deviations were based on conventions in ordinary conversations. Data of the Nebraska Annual Social Indicators Survey 2006 (NASIS) were used for analysis. The results indicate that interviewers tend to shorten lengthy questions, but also add parts to specific questions. In this contribution we will argue that (at least some of) these changes have specific interactional functions: they increased both the cohesion and the coherence within the questionnaire. We will argue that this restructuring of language arises from the fact that an interview is seen as a communicative coherent whole, and not just as a set of individual items. Due to these changes, respondents appeared to be better able to formulate an answer. With the knowledge of this study, the structure within and between questions in surveys can be improved which will enhance response quality.

Introduction

In standardized survey interviews, codified rules are used to minimize measurement errors. Interviewers should read the questions as written, probe for clarification when an inadequate answer is given without pointing the respondent in a direction, collect the answers without changing them for any reason, and be interpersonally nonjudgmental about the content of the answers (Fowler and Mangione, 1990). As Gwartney (2007) argued, the quality of research based on interviews highly depends on the interviewers who collect the data. Nevertheless, when looking at the actual verbal data of an interview, interviewers sometimes deviate from the script when reading questions. According to Ongena and Dijkstra (2006), in many behavior coding studies minor changes by interviewers can increase up to one third of the questions (question meaning not changed) and major changes in up to a quarter of the questions (question meaning changed). Furthermore, interviewer deviations can improve response quality (Dykema, Lepkowski and Blixt, 1997). With this knowledge, we wondered in which way interviewers deviate from questions and how this deviation behavior affected the answering behavior of respondents.

Studies on Interviewer Deviations

Question Rewordings: Respondent Burden or Interviewer Burden?

In explaining measurement error, many researchers have focused on the respondents' mental processes. The model of Tourangeau and Rasinski (1988), in which comprehension, retrieval, judgment and response are the four stages of processing, is mostly used to explain the respondents' cognitive process. According to Krosnick (1991), task difficulty, the ability to perform the task, and motivation are the three factors that cause interceding of this complex mental process in the respondent. Response bias can occur, when a respondent is unable or unwilling to invest enough effort to answer questions.

However, as Ongena and Dijkstra (2007) argued, interviewers can encounter cognitive struggles as well when conducting the interview. In contrast with interviewer effects, such as race and ethnicity (Anderson, Silver and Abramson, 1988), interviewer presence (Schnell and Kreuter, 2005), and interviewer experience (Groves and Couper, 1998), the interviewers' cognitive processes have not been studied extensively in survey research. Schumann and Presser (1997) pointed out that even the smallest changes in question reading affected the respondents' answering behavior. As a matter of fact, standardized surveys may seem simple, but they are not (Clark and Schober, 1992). Many studies have showed that interviewers deviate from the scripted questionnaire (Brenner, 1982; Billiet and Loosveldt, 1988; Schaeffer and Maynard, 1996). Yet, the underlying reason for this interviewer behavior has not received a lot of attention.

Japec (2008) has made an attempt to gain more knowledge about interviewers' mental processes. She based her interviewer process model on the concept of interviewer burden. This concept is defined as "*the total amount of perceived effort, both physical and cognitive, an interviewer has to put in to complete an interview according to specifications*" (Japec, 2008:198). For example, long or tedious interviews, a lot of travel time, and many working hours result in vigorous physical interviewer effort. In addition, the interviewer needs to make a tremendous cognitive effort to interpret confusing instructions, lengthy questionnaires, and understanding difficult questions. According to Japec (2008), measures should be taken to reduce the interviewer burden as this will provide researchers with better data quality.

Ongena and Huiskes (2011) strived to explain interviewer rewordings by using the concepts of interviewer burden as well as respondent burden. Firstly, Ongena and Huiskes argued that deviation behavior was influenced by the interviewers' burden conducting the survey. When questions contained many words or were very complex, for example, Ongena and Huiskes expected that interviewers would shorten the question or change the wording to make their own job easier. Secondly, they argued that interviewer behavior can also be triggered by respondent burden. By shortening questions, using words that are simple to understand, or by providing additional information interviewers want to make the respondents' task easier. Apart from this, interviewers may contextually change the question wordings when they have made certain assumptions about the respondent, e.g. about education and age. However, Ongena and Huiskes' logistic regression analysis showed mixed results. Although there were more deviations in the questions containing difficult words than the ones with easy words, there were fewer reading deviations in longer questions instead of in the shorter questions. In addition, no significant results were found for the respondent variables education and age.

Therefore, Ongena and Huiskes argue the reason for interviewer deviation behavior must be found somewhere else. It is possible interviewers used ordinary conversational techniques when conducting the standardized interview, as Houtkoop-Steenstra found in her studies (2000). These conversational techniques could explain the deviations, and might even improve the quality of the interview. To investigate this, interviewer-respondent interactions should be analyzed with an interactional approach (van't Hof, 2006). This approach is used to examine what occurs on the interviewer-respondent interaction surface.

Rewordings and Interviewer-Respondent Interactions

As Houtkoop-Steenstra (2000) argued, an interview is more than just an easy standard question-answer-acceptance sequence between an interviewer and a respondent. Respondents are used to structures of natural conversations with so-called conversational rules that they come across every day. In ordinary conversation, participants take into account the knowledge of their interlocutors. Hence, a speaker adjusts his language use and topic selection to the recipient, which is called *recipient design* (Sacks, Schegloff and Jefferson, 1974). However, standardized interviews lack this recipient design (Suchman and Jordan, 1990; Houtkoop-Steenstra, 2000), the survey is developed for a large group of respondents. Therefore, the language use and presented topics are the same for everyone.

As respondents have specific linguistic knowledge, they have to invest a lot of effort to participate successfully in a standardized interview. Respondents are not familiar with standardized interviewing rules and roles; they may for example not understand that the answer only needs to contain a number or must be a yes/no reaction. In addition, when questions are asked in natural conversations the addressee knows that the speaker expects an answer after the question has been asked. However, in standardized interviews, interviewers often ask a question whereupon response options follow. Respondents are not always aware of this construction and give an answer before the response options are presented. Furthermore, interviewers may repeat the same response options for several questions in a row as this is scripted. Respondents can find this unusual, and provide the interviewer with an answer before the interviewer has repeated the response options or even interrupt the interviewer while repeating the response options.

Experienced interviewers of course also communicate in non-interview settings. In daily conversations they do not use standardized interviewing rules but the ordinary conversational rules. Therefore, Houtkoop-Steenstra (2000) concluded that interviewer deviation behavior occurs because of the unnatural standardized rules compared to ordinary conversational rules. For example, in ordinary conversations it is common for the speaker to adjust utterances to the answer of the addressee. However, in interviews there is not much room for interviewers to take into account previous answers of respondents because of the standardized rules. This discrepancy causes deviations in question reading as showed in example 1. When a respondent adds extra information to his answer, which already answers an upcoming question, the interviewer still needs to ask the upcoming question as scripted.

Example 1.
Respondent adds extra information to answer (I= interviewer, R= respondent)
I: Do you have a partner?
R: Yes, I live with my husband.
I: And sorry I have to ask this question although you just answered, are you married?
R: Yes, I just told you.

To amplify the assumption that interviewer deviation behavior occurs in standardized interviews because of the differences between conversational- and standardized rules, it is necessary to further study the interview-interactions. People make use of their language knowledge in communicative situations in certain ways (Johnstone, 2008). Information is presented in a certain order, speakers connect word(groups), and try to make the utterance understandable for the addressee. As standardized interviews are not recipient designed but created for all sampled respondents, it is possible the questions do not always match with the individual respondent. As a result, the structure of the survey is less organized.

As Sanchez (1992) argued, the structure of a questionnaire has major influence on the data quality. Therefore, it is important that questions can be linked to each other and are clearly formulated. This clarity in and between sentences is called 'cohesion'. From a discourse perspective, cohesion can be defined as "*the grammatical and lexical relationship within a text or sentence*" (Halliday and Hasan, 1976). Halliday and Hasan (1976) have made a well-known division of five different types of cohesion:

1. Substitution: a word (group) is replaced by another word(group).

   Example 2.
   Substitution (S= speaker, A= addressee)
   S: Have you learned for the exam?
   A: No, I still have to do that.

2. Ellipsis: the omission of words. Participants have to find information elsewhere (e.g. in previous wordings) to understand the context of the conversation.

   Example 3.
   Ellipsis
   S: Who knows where Amsterdam is situated?
   A: I know! (~~where Amsterdam is situated~~)

3. Reference: a word(group) in a sentence refers to another word(group) in the same or another sentence. References can be cataphoric (referring forwards) or anaphoric (referring backwards).

   Example 4a.
   Cataphoric reference
   S: *He* is waiting for the bus. *The professor* had a long way to go.

   Example 4b.
   Anaphoric reference
   S: *The professor* is waiting for the bus. *He* had a long way to go.

4. Conjunction: a word(group) is used to indicate a connection with another word(group). Words as *however*, *but*, and *nevertheless* are such connectives.

   Example 5.
   Conjunction
   S: I usually do not visit the theater, *but* for you I will make an exception.

5. Lexical cohesion: a word(group) is repeated, and a similar or a word(group) that can be related to another word(group) is used.

Example 6.
Lexical cohesion
S: *The university* of Amsterdam has many students. It is a popular *institute.*

In questionnaires, cohesion is concerned with how a word(group) in and between questions is related to another word(group). Often, cohesive structures are implemented in a questionnaire as in example 7.

Example 7.
Cohesion in survey question
I: Out of *four possibilities*, which would be the most important for you in your job?
I: Job at work, adequate salary, personal development, or easy tasks?

Through this cataphoric reference the respondent knows in an early stage that there are four response alternatives he can choose.

Apart from cohesion, the format of a questionnaire should also be coherent.

> "Coherence is a semantic property of discourse formed through the interpretation of each individual sentence relative to the interpretation of other sentences" (van Dijk, 1980).

As Schiffrin (1985) argued, in ordinary conversation people try constantly to make speech coherent. Speakers want their utterances to follow logically on previous utterances, or that their utterances follow logically on utterances of the previous speaker.

Example 8.
Coherence by using numbers
S: *Firstly,* I would like to tell you something about Egypt. *Secondly,* we will discuss queen Cleopatra.

In surveys, coherence is the logical and orderly relation within and between questions in a questionnaire and the place of questions within the whole survey. Most survey designers make an effort to create coherent questionnaires, however as questions are not recipient designed the interviewer sometimes needs to reword questions. For example, an interviewer has to reword the question because of an answer to a previous question. In doing this, the next question follows logically to the previous question.

Example 9.

Coherence in survey question; the respondent already answered the question in a previous question

I: *I know you already told me you are divorced but I still have to ask this question.*

I: What is your marital status?

R: As I said, I am divorced.

We expected that cohesive and coherent cues are also used in standardized interviews, following the argumentation of Houtkoop-Steenstra (2000), because people want to make ordinary conversations cohesive and coherent. Therefore, the research questions this study seeks to answer were:

1. Do interviewer deviations in question reading in standardized interviews contain specific cues of cohesion and/or coherence?
2. If these specific cues are present, how and why are they used?
3. To what extent do interviewer deviations affect the respondent's answering behavior, and possibly as a consequence, improve data quality?

Method

Data Collection

In order to conduct this study, we analyzed interviewer deviations in question reading using the Nebraska Annual Social Indicators Survey 2006 (NASIS). This is an omnibus telephone survey of quality of life among a representative sample of citizens of Nebraska (Bureau of Sociological Research, 2012). The data were collected from November 2006 until March 12, 2007, as part of the NASIS. A random digit dialing (RDD) sampling technique was used, selecting telephone numbers across the State of Nebraska. Respondents were randomly selected from all adults aged 19 or older living in the household. The respondents were interviewed by 44 experienced interviewers from the offices of the Bureau of Sociological Research of the University of Nebraska. Interviews were completed with 1821 respondents. According to AAPOR's (2004) response rate definition 1, the 1821 interviews comprise a response rate of 33%, and a cooperation rate (COOP1) of 42%.

In this survey thirteen questions[1] were included on the topic of Alzheimer's disease. The questions were asked in the second part of the interview, after the interview had already been running for about fifteen minutes, and are preceded by a request to audio record (only this section of) the interview.

Verbal Behavior Coding

The interviews that were recorded (N=300) were transcribed and coded with the Sequence Viewer program (Dijkstra, 2002). Transcripts of the 13 questions on Alzheimer's disease of the interviews were available for sixteen interviewers who completed the interviews. In the 3650 question-answer sequences analyzed, 287 sequences (8%) started with a deviation in question reading.

For 8 of the 13 questions, 1794 question-answer sequences, a deviation in question and/or response alternatives reading occurred in more than 10% of the cases. A coding scheme was designed to capture cohesive and coherent interviewer deviation behavior in these 8 questions (Appendix I and II). Table 11.1 shows the kappa values for the agreement on occurrence categories, and correlations on the agreement on the number of occurrences within an utterance.

Table 11.1: Intercoder reliability: kappa and correlation values

Interviewer verbal behaviors	Kappa values	Correlation values
1. Cohesion		
a. Addition	.734	.720
b. Substitution	.612	.629
c. Deletion	.542	.581
2. Coherence		
a. Addition	.828	.840
b. Addition number (number Q1)	.856	.839
c. Substitution	Not enough cases	Not enough cases
3. Other deviations		
a. Response options not read	.775	.797

The low values found for *deletion cohesion* can be explained by the coders' ability to notice added or substituted word(groups) easier than deleted ones.

Furthermore, to study the effect of interviewer deviation behavior on respondents' answering behavior, a distinction was made between correct answers

1 Some of the questions were manipulated experimentally for purposes of a different study (Ongena, 2009). Due to the very subtle nature of these experimental manipulations, two versions of the same question (questions 1a/1b and 13a/13b) were treated as the same question.

and mismatch answers. A correct answer is defined as a codable answer corresponding to the scripted response alternatives (Ongena, 2005). If a respondent's answer does not correspond with the response alternatives given, the response can be defined as a mismatch answer. However, as example 10 shows, a mismatch answer can still be codable although not exactly corresponding with response alternatives.

Example 10.
Codable mismatch answer question 1 NASIS
I: Which of the following categories would best describe Alzheimer's?
I: Mental illness, neurological disorder, natural effect of aging or viral infection?
R: *The neuro one.*

For this study, we used a strict interpretation for the mismatch answers, namely the answer does not correspond with the given response alternatives and is not codable. In example 11, the respondent chooses two diseases instead of one, therefore the answer is not codable.

Example 11.
Strict mismatch answer question 8 NASIS
I: Which one of the following diseases do you think currently gets the most funding for research and support services?
I: AIDS, Alzheimer's, cancer, diabetes or heart disease?
R: *I think diabetes and AIDS.*

Results

Frequencies

Before turning to the findings of the effects interviewer deviations have on respondents' answering behavior, we looked at the actual occurrences of cohesive and coherent utterances in sequences to find evidence for the assumption that ordinary conversational rules are used in standardized interviews. Table 11.2 provides the frequency of occurrences for each coded category in a sequence. Percentages are also presented.

Table 11.2: Interviewer verbal behaviors: frequencies cohesive and coherent utterances

Interviewer verbal behaviors	Frequency	Percentage
Total of all sequences	1794	100
1. Cohesion		
a. Addition	66	3.68
b. Substitution	122	6.80
c. Deletion	349	19.45
d. Deletion (part of option Q12 and Q13)	25	4.19
Total	562	34.12
2. Coherence		
a. Addition	751	41.86
b. Addition (number Q1)	26	8.70
c. Substitution	4	0.22
Total	781	50.78
3. Other deviations		
a. Response options not read	132	7.36

Of the 1794 sequences coded, we found changes in cohesion of questions in 562 sequences. Interviewers often omitted words, as we found deletions in 349 sequences. In addition, interviewers substituted words in 122 sequences to add cohesion to the question. Less frequently, words were added for cohesive reasons. For questions twelve and thirteen, it was notable that in 25 sequences words were deleted by the interviewers while reading the response options.

Changes in coherence of questions were found in 781 sequences. Remarkably, the category *addition coherence* was coded in 751 utterances. This can be explained by the interviewers' tendency to start their turn with the word 'and' or add 'and' at the starting point of the presentation of the response options. As Table 11.2 shows, *substitution coherence* was only localized in 4 sequences and *deletion coherence* was not found at all. Moreover, we discovered that interviewers added numbers to the response options of question one (a and b) in 26 sequences.

Furthermore, we found that in 132 sequences the interviewers did not read the response options.

Effects on Respondents' Answering Behavior

Cohesion

We found several effects of interviewer deviation on respondents' answering behavior. Table 11.3 shows the cross tabulations of the interviewer's cohesive utterances. For *addition cohesion* there is an association between the cohesive addition utterance and whether an adequate answer was given or not, χ^2 (1, N = 1794) = 4.48, p = .03. Example 12 demonstrates a cataphoric reference constructed with the words 'they are'. Although through the words 'following diseases' the respondent already knew the interviewer was going to present the options later on, the interviewer announced again that the diseases are going to be read by using the words 'they are'.

Example 12.
Addition cohesion question 9 NASIS
I: Which one of the following diseases do you think should get the most funding for research and support services?
I: *They are* AIDS, Alzheimer's, cancer, diabetes or heart disease?

However, looking at the percentages *no addition cohesion* provided fewer mismatch answers than *addition cohesion*. Thus, *addition cohesion* significantly influences the behavior of respondents, but not in a positive manner.

Substitution cohesion on the other hand, has a positive effect on the respondents' answering behavior. Table 11.3 shows that an association was found, χ^2 (1, N = 1794) = 8.79, p = .00, and in terms of percentage more adequate answers were given when a word (group) was substituted. Example 13 demonstrates that the interviewer substituted some words from the original question (Appendix I), he adjusted the question to a preceding answer of the respondent.

Example 13.
Substitution cohesion question 7 NASIS
I: And *is that person* a close family *member*?

The pattern of responses in the substitution conditions is significantly different, *substitution cohesion* significantly influences the answering behaviour of respondents. We further investigated the effects of *substitution cohesion* on question level, but unfortunately there were not enough cases.

No effects were found for deletion cohesion. Although, when analyzing a specific type of the deletion cohesion category, we found positive effects for questions twelve and thirteen. There is an association between the cohesive partial deletion utterance and whether an adequate answer was given or a mismatch answer was provided, χ2 (1, N = 597) = 5.70, p = .01. In questions twelve and thirteen interviewers omitted parts of the response options, which corresponds with the ellipsis and lexical cohesion categories of Halliday and Hasan (1976). Example 14 demonstrates that the interviewer deleted the word familiar from the enumeration.

Table 11.3: Cross tabulations: cohesion

Interviewer verbal behaviors				
		No Addition	Addition	Total
a. Addition	Adequate answer	1494 (86.46%)	51 (77.27%)	1545
	Mismatch answer	234 (13.54%)	15 (22.73%)	249
	Total	1728	66	1794
	Chi Square χ^2	(1, N = 1794) = 4.48, p = .03		
b. Substitution		No Substitution	Substitution	Total
	Adequate answer	1429 (85.47%)	116 (95.08%)	1545
	Mismatch answer	243 (14.53%)	6 (4.92%)	249
	Total	1672	122	1794
	Chi Square χ^2	(1, N = 1794) = 8.79, p = .00		
c. Deletion		No Deletion	Deletion	Total
	Adequate answer	1247 (86.30%)	298 (85.39%)	1545
	Mismatch answer	198 (13.70%)	51 (14.61%)	249
	Total	1445	349	1794
	Chi Square χ^2	(1, N = 1794) = 0.19, p = .65		
d. Deletion (part of option Q12 and Q13)		No Deletion (part of option)	Deletion (part of option)	Total
	Adequate answer	430 (75.17%)	24 (96%)	454
	Mismatch answer	142 (24.83%)	1 (4%)	143
	Total	572	25	597
	Chi Square χ^2	(1, N = 597) = 5.70, p = .01		

Example 14.

Deletion part of response option question 12 NASIS (uuh = hesitation marker)

I: And how familiar are you with the services provided by the Alzheimer’s Association?

I: Uuh very familiar, familiar, somewhat ~~familiar~~, not very familiar?

Table 11.3 shows that there are significantly different response patterns in the partial deletion conditions. Considering the percentages, partial deletion of response options in questions twelve and thirteen has a positive effect on the respondents’ answering behavior since fewer mismatch answers occurred.

Coherence

Table 11.4 shows the cross tabulations of the interviewers’ changes in coherence of questions. Even though *addition coherence* was coded in 751 sequences, no effects on respondents’ answering behaviour were found for this category. Example 15 demonstrates which words were added by the interviewer. The word ‘and’ is used to connect questions, and the words ‘would you say’ are added as a signal for the respondent that he is supposed to give an answer after the response options are presented.

Example 15.

Addition coherence question 8 NASIS (um = hesitation marker)

I: *And* which one of the following diseases do you think currently gets the most funding for research and support services?

I: Um *would you say* AIDS, Alzheimer's, cancer, diabetes or heart disease?

Interpreting the percentages, it seems that *addition coherence* has a somewhat more negative effect on respondents’ answering behaviour than when *addition coherence* was not used.

However, we found that *addition coherence* sometimes has positive effects on the response quality of respondents. Table 11.4 shows that there is an association between the coherent numbered addition utterance and whether an adequate answer was given or not, χ^2 (1, N = 299) = 10.76, p =. 00. When interviewers added numbers to the response options in question 1 as example 16 demonstrates, respondents were better able to formulate an answer.

Table 11.4: Cross tabulations: coherence

Interviewer verbal behaviors				
		No Addition	Addition	Total
a. Addition	Adequate answer	907 (86.96%)	638 (84.95%)	1545
	Mismatch answer	136 (13.04%)	113 (15.05%)	249
	Total	1043	751	1794
	Chi Square χ^2	$(1, N = 1794) = 1.47$, $p = .22$		
b. Addition (number Q1)		No Addition (number)	Addition (number)	Total
	Adequate answer	191 (69.96%)	26 (100.00%)	217
	Mismatch answer	82 (30.04%)	0 (0.00 %)	82
	Total	273	26	299
	Chi Square χ^2	$(1, N = 299) = 10.76$, $p = .00$		
c. Substitution	Not enough cases			

Table 11.5: Cross tabulations: response options not read

Interviewer verbal behaviors				
		Read (no deletion)	Not read (deletion)	Total
a. Response options not read	Adequate answer	1490 (89.65%)	55 (41.67%)	1545
	Mismatch answer	172 (10.35%)	77 (58.33%)	249
	Total	1662	132	1794
	Chi Square χ^2	$(1, N = 1794) = 235.55$, $p = .00$		

Adding numbers to questions led to a more coherent structure of the question.

Example 16.

Addition of numbers question 1 NASIS

I: Which of the following categories would best describe Alzheimer's?

I: *One*, Mental illness, *two*, neurological disorder, *three*, natural effect of aging or *four*, viral infection?

Notably, in all the sequences interviewers added numbers an adequate answer was given by the respondents.

Unfortunately, there were not enough cases to analyze the significance of effects of *substitution coherence* on the answering behaviour of respondents.

Other Deviations

In 132 sequences interviewers did not present the response options to the respondent. We found an association between the deletion of response options and whether an adequate answer was given or a mismatch answer was provided, χ^2 $(1, N = 1794) = 235.55, p = .00$.

Example 17.

Response options not read question 13 NASIS (uuh = hesitation marker)

I: Uuh, if someone in your family or yourself developed Alzheimer's disease, how likely would you be to turn to the Alzheimer's Association?

The statistical analyses and the percentages presented in table 11.5 both display that the deletion of response options, as demonstrated in example 17, significantly influences the behavior of respondents negatively since much more mismatch answers were found.

Discussion

Our study confirms the appearance of conversational principles in standardized interviews and highlights the importance of studying interviewer-respondent interactions. We found that many deviations made by interviewers have specific interactional functions: changes in question reading are made to improve cohesion and coherence in the survey. We argue that these deviations arise from the fact that an interview is seen as a communicative coherent whole, and not just as a set of individual items.

Although the changes on first appearance do not seem to change the meaning of the questions, standardized interviewing rules exist for a reason, as we found that *addition cohesion* negatively influenced the respondents' answering behavior. However, question rewording does not necessarily decrease response quality. More adequate answers were given when interviewers substituted word(groups), partially deleted words when presenting response options in questions twelve and thirteen, and added numbers to the response options in question one. Due to these changes, respondents appeared to be better able to formulate an answer.

In the following examples we proposed some improvements for the NASIS survey to enhance response quality in telephone interviews. For a start, according to our results adding numbers to questions with categorical response alternatives for coherent purposes as in example 18, have a positive effect on respondents' answering behavior.

Example 18.

Addition of numbers for coherent structure question 1 NASIS

I: Which of the following categories would best describe Alzheimer's?

I: *One*, mental illness, *two*, neurological disorder, *three*, natural effect of aging or *four*, viral infection?

Furthermore, survey designers should notice that questions in a survey should always correspond to previous answers of respondents. For example, questions that can be answered with a singular (e.g. one or person) or plural response (two, three etc. or people) or questions in which answers with different verb tenses are imaginable (e.g. I have or I had) require different follow-up questions. When interviewers are not provided with corresponding follow-up questions, they have to deviate from the script to create a cohesive structure. Taking this into account, an alternative question should be formulated for the interviewer as is implemented in example 19.

Example 19.

Alternative question for cohesive structure question 6 and 7 NASIS

Q6 I: What is the number of people you know personally, if any, who have Alzheimer's disease?

Singular answer: question 7a / Plural answer: question 7b

Q7a I: Is this person with Alzheimer's disease a close family member?

Q7b I: Are any of these people with Alzheimer's disease close family members?

As all other scientific studies, this research has its imperfections. First of all, more data should be collected to expand the amount of sequences to support the results presented in this study. Secondly, since the respondents were questioned on a sensitive topic (Alzheimer's disease) a replication of the analyses on non-sensitive questions is advised. In addition, experimental studies are necessary to further investigate the effects of interviewer deviations on respondent answering behavior. It is possible that certain interviewer effects influenced the respondent and that the adding of numbers to response options is not the only reason for respondent's answering behavior.

Ultimately, with the knowledge of this study and essential future research the structure within and between questions in surveys can be improved which will enhance response quality.

Bibliography

Anderson, Barbara A., Silver, Brian D. and Abramson, Paul R.: The effect of race of the interviewer on race-related attitudes of black respondents in SRC/CS national election studies, Public Opinion Quarterly, 52, 289-324, 1988.

Billiet, Jacques and Loosveldt, Geert: Improvement of the quality or responses to factual survey questions by interviewer training, Public Opinion Quarterly, 52, 190-211, 1988.

Brenner, Michael: Response-effects of "role-restricted" characteristics of the interviewer, In: Dijkstra, Wil and van der Zouwen, Johannes (eds.): Response Behavior in the Survey Interview (131-165), London: Academic Press, 1982.

Bureau of Sociological Research, Nebraska, Annual Social Indicators Survey, 2012. Available at: http://bosr.unl.edu/nasis

Clark, Herbert H. and Schober, Michael F.: Asking questions and influencing answers, In: Tanur, Judith M. (ed.): Questions About Questions: Inquiries Into the Cognitive Bases of Surveys (15-48), New York: Sage, 1992.

Dijkstra, Wil: Transcribing, coding, and analyzing verbal interactions in survey interviews, In: Maynard, Douglas W., Houtkoop-Steenstra, Hanneke, Schaeffer, Nora C. and van der Zouwen, Johannes (eds): Standardization and Tacit Knowledge: Interaction and Practice in the Survey Interview (401-425), New York: Wiley, 2002.

Dykema, Jennifer, Lepkowski, James M. and Blixt, Steven: The effect of interviewer and respondent behavior on data quality: analysis of interaction coding in a validation study, In: Lyberg, Lars E., Biemer, Paul, Collins, Martin, De Leeuw, Edith D., Dippo, Cathryn, Schwarz, Norbert and Trewin, Dennis (eds.): Survey Measurement and Process Quality (287-308), New York: Wiley, 1997.

Fowler, Floyd J. and Mangione, Thomas W.: Standardized Survey Interviewing: Minimizing Interviewer-related Error, Newbury Park, Sage, 1990.

Groves, Robert M. and Couper, Mick P.: Nonresponse in Household Interview Surveys, New York NY, Wiley, 1998.

Gwartney, Patricia A.: The Telephone Interviewer's Handbook, San Francisco, Jossey-Bass, 2007.

Halliday, Michael A.K. and Hasan, Rugaiya: Cohesion in English, London, Longman, 1976.

Houtkoop-Steenstra, Hanneke: Interaction and the Standardized Survey Interview. The Living Questionnaire, Cambridge, University Press, 2000.

Japec, Lilli: Interviewer error and interviewer burden, In: Lepkowski, James M., Tucker, Clyde, Brick, J. Michael, de Leeuw, Edith D., Japec, Lilli, Lavrakas, Paul J., Link, Michael W. and Sangster, Roberta L. (eds.): Advances in Telephone Survey Methodology (187-211), Hoboken NJ: John Wiley & Sons, 2008.

Johnstone, Barbara: Discourse Analysis, Oxford, Blackwell Publishing, 2008.

Krosnick, Jon A.: Response strategies for coping with the cognitive demands of attitude measures in surveys, Applied Cognitive Psychology, 5, 213-236, 1991.

Ongena, Yfke P.: Helping respondents to format their answers: a question wording experiment in a telephone survey, Presented on 64th Annual Conference of AAPOR, Hollywood FL, United States, 2009.

Ongena, Yfke P.: Interviewer and Respondent Interaction in Survey Interviews, Ph. D. Dissertation, University of Amsterdam, Netherlands, 2005.

Ongena, Yfke P. and Dijkstra, Wil: A model of cognitive processes and conversational principles in survey interview interaction, Applied Cognitive Psychology, 21, 145-163, 2007.

Ongena, Yfke P. and Dijkstra, Wil: Methods of behavior coding of survey interviews, Journal of Official Statistics, 22, 419-451, 2006.

Ongena, Yfke P. and Huiskes, Mike: Interviewer's rewording of questions in CATI surveys, Presented on 66th Annual Conference of AAPOR, Phoenix AZ, United States, 2011.

Sacks, Harvey, Schegloff, Emanuel A. and Jefferson, Gail: A simplest systematics for the organization of turn taking in conversation, Language, 50, 696-735, 1974.

Sanchez, Maria E.: Effects of questionnaire design on the quality of survey data, Public Opinion Quarterly, 56, 206-217, 1992.

Schaeffer, Nora C. and Maynard, Douglas W.: From paradigm to prototype and back again: interactive aspects of "cognitive processing" in standardized survey interviews, In: Schwarz, Norbert and Sudman, Seymour (eds.): Answering Questions: Methodology for Determining Cognitive and Communicative Processes in Survey Research (65-88), San Francisco: Jossey-Bass, 1996.

Schiffrin, Deborah: Conversational coherence: the role of well, Language, 61, 640-667, 1985.

Schnell, Rainer and Kreuter, Frauke: Separating interviewer and sampling point-effects, Journal of Official Statistics, 21, 389-410, 2005.

Schumann, Howard and Presser, Stanley: Questions and Answers in Attitude Surveys: Experiments on Question Form, Wording, and Context, New York, Academic Press, 1981.

Suchman, Lucy and Jordan, Brigitte: Interactional troubles in face-to-face interviews, Journal of the American Statistical Association, 85, 232-241, 1990.

Tourangeau, Roger and Rasinski, Kenneth A.: Cognitive processes underlying context effects in attitude measurement, Psychological Bulletin, 3, 299-314, 1988.

van Dijk, Teun A.: Macrostructures: an Interdisciplinary Study of Global Structures in Discourse, Interaction, and Cognition, Hillsdale NJ, Lawrence Erlbaum, 1980.

van 't Hof, Sanne: From Text to Talk. Answers and Their Uptake in Standardized Survey Interviews, Ph. D. Dissertation, University of Utrecht, Netherlands, 2006.

Appendix

I. NASIS Survey Instrument (2006)

Bureau of Sociological Research, University of Nebraska - Lincoln

Question 1a

Which of the following categories would best describe Alzheimer's?
Mental illness, neurological disorder, natural effect of aging or viral infection?

Question 1b

What would be the best way to describe Alzheimer's?
Mental illness, neurological disorder, natural effect of aging or viral infection?

Question 7

Are any of these people with Alzheimer's disease close family members?

Question 8

Which one of the following diseases do you think currently gets the most funding for research and support services?
AIDS, Alzheimer's, cancer, diabetes or heart disease?

Question 9

Which one of the following diseases do you think should get the most funding for research and support services?
AIDS, Alzheimer's, cancer, diabetes or heart disease?

Question 12

How familiar are you with the services provided by the Alzheimer's Association?
Very familiar, familiar, somewhat familiar, or not very familiar?

Question 13a

If someone in your family or yourself developed Alzheimer's disease, how likely would you be to turn to the Alzheimer's Association for information?
Very likely, somewhat likely, or not at all likely?

Question 13b:

If someone in your family or yourself developed Alzheimer's disease, how possible is it that you would turn to the Alzheimer's Association for information? Is there a…
(Is there a) good possibility, a small possibility or no possibility at all?

II CODING SCHEME

Interviewer verbal behaviors	**Descriptions**
Main example	Question 9 NASIS used as example: *I: Which one of the following diseases do you think should get the most funding for research and support services?* *I: AIDS, Alzheimer's, cancer, diabetes or heart disease?*
1. Cohesion	
a. Addition	Interviewer adds words that can be marked as added cohesion in a sentence. *I: Which one of the following diseases do you think should get the most funding for research and support services?* *I: They are AIDS, Alzheimer's, cancer, diabetes or heart disease?*
b. Substitution	Interviewer substitutes words that can be marked as substituted cohesion in a sentence. *I: Which one of the following diseases do you think should get the most funding for research and support services?* *I: They are the same five as question 8*
c. Deletion	Interviewer deletes words that can be marked as deleted cohesion in a sentence. *I: Which ~~one~~ of the following diseases do you think should get the most funding for research and support services?* *I: AIDS, Alzheimer's, cancer, diabetes or heart disease?*
d. Deletion (part of option Q12 and Q13)	Interviewer deletes part of response options that can be marked as deleted cohesion in a sentence. Question 12 NASIS used as example: *I: How familiar are you with the services provided by the Alzheimer's Association?* *I: Very familiar, familiar, somewhat ~~familiar~~, or not very familiar?*
2. Coherence	
a. Addition	Interviewer adds words that can be marked as added coherence in a sentence. *I: and which one of the following diseases do you think currently gets the most funding for research and support services?* *I: AIDS, Alzheimer's, cancer, diabetes or heart disease?*

Interviewer verbal behaviors	**Descriptions**
b. Addition (number Q1)	Interviewer adds numbers to response options that can be marked as added coherence in a sentence. Question 1a NASIS used as example: *I: Which of the following categories would best describe Alzheimer's?* *I: One, Mental illness, two, neurological disorder, three, natural effect of aging or four, viral infection?*
c. Substitution	Interviewer substitutes words that can be marked as substituted coherence in a sentence. *I: Which one of the following diseases do you think currently gets the most funding for research and support services?* *I: AIDS, Alzheimer's, cancer, diabetes and heart disease?*
3. Other deviations	
a. Response options not read	Interviewer does not read response options. *I: Which one of the following diseases do you think currently gets the most funding for research and support services?* *R: AIDS*

Mischief versus Mistakes: Motivating Interviewers to Not Deviate

Patricia A. Gwartney

Abstract

This chapter identifies patterns of interviewer mischief (deliberate falsification) and interviewer mistakes (accidental error) in surveys. Using the theoretical concepts of "total error" in the life cycle of a survey, the author outlines how all participants in the survey process can err and thereby contribute to total error, but interviewers' role is unique because they are the link between respondents and all survey science. They can make or break researchers' ability to test hypotheses, develop new theory, and evaluate social policies and programs. Despite their importance, interviewers remain unseen and under-valued, with perhaps lowest-prestige jobs in survey organizations. The author details why and how interviewers may falsify or misreport survey data, and presents techniques to detect and prevent such acts. Ultimately, this chapter argues that the key to minimizing rare interviewer mischief, as well as mistakes, is for survey research organizations create an employment culture that recognizes and rewards interviewers' crucial role in the data collection process. Data for this essay stems from years of participant observation at all levels of the survey process.

Introduction

In this information age, survey data permeate multiple dimensions of public life and frame society's perception of itself. Survey researchers must do everything in their power to assure public trust in the quality of evidence based on survey data collection. Interviewer falsification of data, the focus of this volume, involves intentional deviations from acceptable practices and survey instructions as the interviewer delivers the survey instrument. Preventing interviewer-related mischief is essential to maintaining our implied pact with the public.

Interviewer falsification is not equivalent to interviewer error. Interviewer error involves unintentional mistakes that are thought to be random, not systematic. Interviewer deviance is systematic and can result in biased and misleading survey results. The distinction between mischief and mistakes is important for survey research organizations (SROs) because the latter is correctable, whereas the former suggests a poisonous employment situation.

The potential for systematic interviewer error has been recognized for decades by SROs, resulting in rigorous interviewer training protocols to standardize interview delivery (Crespi 1945, Gwartney 2007). The potential for interviewer falsification has also been recognized for decades, resulting in SROs' developing codes of interviewer ethics to avoid cheating and instituting strict, routine practices to discover it should it occur. While a handful of legendary stories of interviewer falsification have circulated among SROs in North America for years, the pervasiveness of interviewer-counterfeited data appears to be low (American Association for Public Opinion Research 2009), indeed as low as one tenth of one percent in a large Canadian survey (Murphy, Eyerman, McCue, Hottinger, and Kennet 2005).

Practices to prevent falsification differ by type of survey. Interviewer-administered surveys occur in two environments: (1) centralized telephone facilities with high degrees of supervision and (2) dispersed field situations for face-to-face interviewing with remote supervisor contact. The potential for mischief also exists when interviewers are involved in dropping off or picking up self-administered questionnaires. Preventative systems differ for these situations.

While some chapters in this volume are dedicated to statistical procedures designed to mechanically detect interviewer data falsification, I advocate for a more organic approach. Specifically, I build the argument that SROs must recognize and value interviewers' unique contributions, reward them appropriately, and create an organizational culture in which interviewer deviance is itself deviant.

I start by contextualizing interviewer deviance in the life cycle of the survey and employing the concept of "total survey error" (Andersen, Kasper, and Frankel 1979; Biemer 2010; Weisberg 2005). I identify the points in the survey process where interviewers can deviate in the process of doing their job, using examples to distinguish between interviewer mistakes and outright falsification.

Interviewer Deviance in the Life Cycle of the Survey

The processes of developing and conducting a survey are complex but follow predictable steps from start to finish. In addition to the interviewing crew, they typically involve an array of scientists, research assistants, programmers, field managers, budget managers, coders and clerks. Any of these persons can deviate intentionally from protocols or err accidentally. This section locates interviewers in the life cycle of the survey as a way of creating a context to understand interviewers' contribution to survey error.

The sequence of the survey process begins by identifying a research question, defining hypotheses and concepts, planning a budget and timeline, seeking funding through lengthy peer reviews, then operationalizing concepts into survey questions, designing the survey instrument, and an iterative process of pretesting and revision. Simultaneously, respondent email or postal communications are developed and tested.

As the project advances, another crew works on the sample, i.e., identifying the target population, finding or creating a list to serve as the sample frame, implementing a randomization process to select units, and developing methods for interviewers to use sample selection procedures in the field.

Next is hiring and then training interviewers–training not just to read questions but to understand the concepts behind the questions, how and when to use probes and alternative phrases, how to select respondents, and, importantly, how to use their computer hardware and software.

Once the SRO looses interviewers into the field, they must be monitored and supervised, updates and revisions must be communicated, equipment must be maintained, the sample must be monitored, data must be uploaded and downloaded, and problems must be resolved quickly or precious interviews may be lost. After the easy respondents are picked off, interviewers must track down those who are hard to reach and convince those who initially refused to participate.

Eventually, data collection ends, the processes of data reduction and data analysis commence, and reports are written. The entire process may last six weeks or six years and can involve hundreds of people acting in various roles.

Mistakes and mischief can occur in any of these steps in the life of a survey, by interviewers and non-interviewers alike, contributing to error and potential bias. For example, principle investigators may overlook a key concept or employ someone else's without attribution. Respondents can misreport their behaviors because they do not recall accurately or because they do not want to reveal their true behaviors. Data analysts can err in sorting datasets and in creating scales and indexes, or overlook results inconsistent with their hypotheses.

Survey methodologists divide survey error into two components: that due to sample selection, called sampling error, and that due to everything else, called nonsampling error. Together they comprise "total survey error" (Andersen, et al. 1979; Biemer 2010; Weisberg 2005). The concept of total error redirected researchers' fixation on sampling error to considering the potential for errors in all aspects of the survey process, including that attributable to interviewers. Still, the methods to measure nonsampling error remain under developed. The point is that the survey life cycle creates the potential for more actors' deviations than any attributable to interviewers alone.

Yet we focus on interviewers because they remain the link between the SRO and the respondent. Interviewers are the face of the SROs employing them. Interviewers can make or break researchers' ability to test hypotheses, develop new theory, and create social policies and programs. Most of the general public encounters social science only when talking with interviewers. For these reasons, interviewer mischief carries disproportionate weight compared to the other actors involved in the multitudinous tasks of the survey process.

Where in the Survey Life Cycle Interviewer Deviance Can Occur

It is important to understand that interviewers who want to cheat can do more than falsify interviews. In the course of performing their job, interviewers can deviate from acceptable performance or procedure at the time of initial contact when respondents are chosen and when administering the survey instrument and recording data. Below I briefly discuss each.

Initial Respondent Contact and Sampling

In their initial interactions with potential respondents, the interviewer's job involves introducing the study, explaining the subject matter and the probable length of the task, assuring respondents of confidentiality, addressing their concerns and questions, and implementing the sample design to select the intended respondent. The biggest problems occurring at this stage include interviewers selecting the wrong element for the sample, interviewing the wrong individual, or allowing a potential respondent to refuse.

How can interviewers interview the wrong person? The first person with whom the interviewer speaks may not be the intended respondent; in this instance, the interviewer often must gain the cooperation of the initial contact in order to implement a sampling procedure that identifies the intended respondent (e.g., a household listing to randomly select an adult). If the intended respondent is available, the interviewer repeats the study explanation, answers that person's questions, and the interview proceeds. If that person is not available, the interviewer arranges to re-contact the household at a time when the intended respondent is likely to be present. Problems arise when the intended respondent is difficult to locate or the initial contact acts as a gatekeeper, refusing access to the intended respondent; both complicate the interviewers' job and lengthen the time to complete the interview.

Interviewers who are easily frustrated, discouraged by repeatedly unsuccessful callbacks, attempting to meet a quota, or fear performing refusal conversions may be tempted to record their interactions with the initial respondent incorrectly. For example, rather than call back at a household where a protective wife disallows access to her busy husband, such an interviewer may be tempted to record the dwelling as vacant or, in the case of a telephone interview, as a non-working number, so that no interviewer will have to call back again. If the wife of the busy spouse volunteers to be interviewed in his stead, a weak-willed interviewer may accept her as a respondent.

On the other hand, even motivated interviewers sometimes err accidentally, e.g., misinterpret an address, get lost in a new area, misread a previous interviewer's notes about the case, or mis-key a telephone number. Even experienced interviewers sometimes have incomplete skill sets. For example, because encounters with gatekeepers – or, worse, bullies – are infrequent, even interviewers fully trained in handling intimidation may not have had the opportunity to practice those skills enough to use them effectively, causing them to make mistakes in following the sampling protocol.

Consider the following real example: A motivated but newly launched interviewer knocked on the door of a friendly older couple's home. The newly retired husband answered the initial questions but the sampling instructions identified his wife as the intended interviewee. Both spouses are unhappy with this, for the survey's subject matter (transportation) involved his daily experience more than hers. The interviewer guided the wife to sit in the dining room, apart from where the husband watched television in the living room. But he eavesdropped; he interjected his opinion; he gradually slid his chair closer and closer; and he eventually took over the interview from his wife, who slipped away with evident relief. She brought back homemade cookies. The interviewer's objections were insufficiently strong to overcome this couple's gendered division of labor. The SRO sent an experienced interviewer the next day to re-do the interview. The interviewer could have covered her faux pas easily, for she delivered a completed interview. What was it about her training or SRO that made her tell the truth?

Telephone interviewers have talked frankly with me about the difficulties of recontacting households whose dial attempt records show multiple "soft" refusals or "phone slams." They admitted the temptation to falsify call records in order to eliminate difficult cases. Why didn't they do it? What prevented their deviance was the ability to explain the circumstances of their lack of success in computer-aided telephone interview software and, on the flip side, the opportunity to read previous interviewers' detailed reports which helped them to develop a callback strategy.

No matter if the problem was accidental or deliberate, interviewers who do not consistently follow sample selection instructions contribute to sampling error and restrict their study's generalizability. Interviewers who allow respondents to refuse contribute to coverage error. Each person, household, or organization intended for a sample represents thousands of others like them. If they are not represented in the survey results, the study's results may be biased and misleading, particularly if the omitted units share reasons for not participating. Interviewers must learn to effectively use both refusal conversion techniques and their personal powers of persuasion in order to avoid unit-level refusals. Properly trained and motivated interviewers should never be a source of sampling error or coverage error.

Administering the Instrument and Recording Answers

Once the correct respondent is selected and ready to participate, it is the interviewer's job to administer the instrument using standardized interviewing techniques combined with their knowledge of the psychology of how respondents answer questions (Gwartney 2007, Tourangeau, Rips, and Rasinski 2000). In brief, they read all questions exactly as they are written, in the order presented, in a neutral tone of voice, and offer neutral feedback at appropriate intervals. Interviewers should never utter a word or sound or otherwise behave in a manner that could influence or change a respondent's answers. They should not suggest answers, ask unscripted questions, comment on respondents' answers, or try to make respondents' answers consistent across questions.

Theoretically, uniform survey administration maximizes the validity and reliability of data for subsequent analysis by eliminating any variation in respondents' answers that could be attributed to interviewers. However, uniform administration is not the same as rote recitation. Instead, questionnaire designers employ the tacit knowledge interviewers develop with training and experience to systematize normal conversational variation. On the basis of actively listening, interviewers use scripted "fill" to tailor the conversational flow of the interview to individual respondents without sacrificing data quality (Gwartney 2007, Dillman, Smyth, and Christian 2009).

"Fill" refers to a consistent set of structured variations in question wording written into the survey instrument. Employing fill in a methodical manner builds coherence into the conversational structure of interviews (Maynard, Houtkoop-Steenstra, Schaeffer, and van der Zouwen 2002). Interviewers use their tacit knowledge to systematize normal language variations in their choice of scripted nondirective probes, both to prevent item non-response and to improve respondents' interview experience. Standardized interviewing techniques also convey professionalism which puts respondents at ease.

To implement a standardized conversation, it is helpful to think of the four essential steps involved in the question-answer sequence: (1) The interviewer reads a question, (2) the respondent answers, (3) the interviewer acknowledges that answer, and then (4) records it. For example,

(1) Interviewer: "How many years have you lived in [city]?"
(2) Respondent: "Five years."
(3) Interviewer: "Thank you."
(4) Simultaneously, the interviewer types "5."

Deviations in the question-and-answer sequence can cause interviewers to make mistakes Examples of deviations in step 1 include an interviewer allowing a respondent to interrupt before the question is read completely or an interviewer suggesting a specific answer to a hesitant respondent. Problems can occur in step 2 when a respondent gives a nonconforming answer (e.g., "I moved in six months ago"), says something that does not answer the question (e.g., "My mother moved here six years ago"), or elaborates beyond a reasonable answer in a manner that contradicts it ("Five years, but my mother lived here a year before that. She kicked me out two years ago but let me come back later").

Step 3 above is a form of feedback. As much as interviewers are highly trained in administering surveys, respondents are not; i.e., the persons randomly selected to participate in survey interviews rarely have done so before and do not automatically know how to behave as good respondents. Well-trained interviewers evaluate the quality of respondents' answers and employ positive and negative feedback phrases that implicitly guide respondents into performing their role well in the interaction.

Properly trained interviewers probe inadequate answers in order to arrive at an appropriate one. However, an interview suffused with elaborations can motivate an interviewer to innovate (Maynard and Schaeffer 2006). Suppose, for example, a respondent gives a long-winded answer to an open-ended question at step 2 but the exasperated interviewer records only the first 20 words at step 4. Is such deliberate deviation from protocol an interviewer error of omission, or is it a form of falsification?

Another example stems from interviewers learning things about respondents indirectly, e.g., when elaborations to one question provide information that directly answers a subsequent question. An interviewer who uses that information to skip step 1 and directly record an answer in step 4 has clearly falsified data. Less obvious is the case in which an interviewer records an "other-specify" answer category in step 4 when a simple, scripted probe would have resulted in more usable data.

Interviewers may be tempted to employ shortcuts to speed a difficult or uncomfortable interview toward its conclusion. The failure to probe, tailor, or guide gently with feedback can make interviews sound perfunctory to respondents and convey a lack of care or seriousness about the task, which in turn will not inspire respondents to answer fully or with care. Such data collection problems occur even among well-trained interviewers if they are sick, afraid, paid by the piece, or feel alienated from their employer. These shortcomings in delivering the surveys are probably best labelled as interviewer error, not falsification, even if the interviewer acted deliberately.

Item non-response (i.e., answers such as "don't know," "refuse," "other," "no opinion," "no answer," and "if volunteered") can undermine a study's validity and reliability. Non-responses are non-answers; they leave researchers without substantive data to analyze for those cases. Interviewers should be fluent in the practices that discourage item non-response, unless specifically instructed otherwise (e.g., knowledge questions with a "don't know" option). Respondents most often refuse to answer questions with sensitive content, i.e., those related to sexuality, religion, illegal behavior, and drug/alcohol use, as well as questions about socially desirable behavior, such as voting, volunteering, exercise, and giving to charity. Interviewers should know how to encourage respondents to select from the answer categories provided, thereby avoiding non-response. When interviewers fail to use those techniques, are they falsifying the data or is it a form of interviewer error?

Importantly, data collection errors can occur among interviewers eager to please as well as among those less motivated. For example, I witnessed an overly thorough interviewer continue to repeat the common answer categories for a question series long after her respondent should have remembered them. Few respondents need to hear "do you strongly agree, somewhat agree, somewhat disagree, or strongly disagree" 10 times for 12 questions; most respondents learn them after just two or three repetitions. Similarly, some interviewers continue to probe even after respondents have given complete responses. These practices lengthen interviews and try respondents' patience, which can later result in cut-offs and incomplete answers.

Skip logic is rarely a source of error any longer, since most modern interviewer-administered surveys are computerized and the software is programmed to omit inappropriate questions, such as job-related questions for unemployed respondents. Before the days of computer software, such data collection errors sometimes occurred, but it is hard to imagine an interviewer deliberately reading inappropriate questions, e.g., asking teenagers about their grandchildren. On the other hand, dedicated cheaters can use skip logic to manipulate the length of a survey to a minimum (Brüderl, Huyer-May, Schmiedeberg 2013, this volume), e.g., guiding respondents with two part-time jobs through the employment questions once instead of twice. I address such interviewers in the next section.

Why Some Interviewers Falsify Data and How They Do It

This section examines types of interviewers who falsify data, factors contributing to the decision to falsify, and patterns of falsification.

Calculating Cynics, Striving Incompetents, and Temporary Ethical Breakdowns

Why do some interviewers fake data? Among the plethora of individual and organizational factors that can interact to cause lying, cheating and mistakes, three types of interviewers emerge. At one extreme is the "calculating cynic", whose cost-benefit analysis shows that faking interviews is easier than doing the job correctly and who pursues falsification with diligence. At the other extreme is the "striving incompetent," who really wants to perform well but errs repeatedly, despite coaching and re-training. In between are apt, hardworking interviewers whose ethics occasionally break down in difficult or unusual situations.

To the calculating cynic faking interviews is a means of completing one's work in the least amount of time with the least amount of effort and still receive the maximum possible pay. From this perspective, rational interviewers simplify their work lives by faking all or part of their completed interviews.

But how is it possible to fake interviews without getting caught? Blasius (2011) provided a "user's guide" for how face-to-face interviewers can optimize pay while minimizing effort. The advice included: Keep a copy of each cooperative respondent's address; work for multiple SROs in order to return to them multiple times. Prepare a list of friends and friends of friends that you can use over and over as respondents (or confederate respondents), especially for surveys with common characteristics. Find other faker interviewers and share lists with them, but be sure they are the same sex and about the same age in case your employer validates completed interviews by recontacting respondents. Create logical and plausible answers to the survey questions, but not elaborate ones. Reduce the instrument to the bare minimum; attend to the skip logic to strip out the most questions. Make sure faked respondents match the socioeconomic, demographic, and political characteristics of the target population, and map their characteristics onto appropriate neighborhoods.

To interviewers who take Blasius' advice and try to become master fakers of interviews: Does it matter to you that keeping identifying information on respondents, like addresses, violates human subjects' protections? How long do you think you will keep those "friends" that you and your fellow deviants keep interviewing over and over? Do you really think the compliant respondents will not remember you from one or more prior interviews? Consider creating a spreadsheet of who you use for what survey and when, so that you do not accidentally return to them too soon. Better, pretend that they are part of a panel study to give you an excuse to keep returning. If your employer is likely to use your sex and age to identify you as someone's interviewer, when you recruit fellow deviants be sure they look similar to you or are willing to modify their looks (height, weight, eye color, hair color and length, style of dress).

To create logical and plausible answers to the survey questions, it would be best to create fictitious but internally consistent types of individuals, e.g., the striving young immigrant, the damaged grandmother, the angry man of color, the rigidly aloof professional, and the conventional suburban mother. Do not forget that a certain percent of your respondents must be Democrats and a certain percent Republicans, and be sure to map those fake Republicans into the higher-end neighborhoods. If the instrument asks about products that you do not own, go out and buy them or do your homework (google the item or get to the library to consult an encyclopedia) to ensure your faked answers do not sound too counterfeit. Be sure to keep copies of your faked interviews for the sake of future consistency.

Is this carefully crafted and calculated behavior actually rational? Interviewers who follow this advice are likely to end up both friendless and jobless, and possibly facing civil litigation for violating their oath to protect human subjects. Consider the hours, weeks, or even months it will take to create fictitious persons, research population demographics and map them, research products, round up other interviewers who look like you, and enlist respondents willing to go along with your ruse.

Consistently faking interviews, in fact, seems a lot more time consuming and more complicated than the work "irrational" interviewers perform as they plod through real people's homes obtaining authentic data. In addition, persistent interview fakers are likely to experience some psychic cost that is hard to measure (Eisenhauer 2006). The psychic costs of living a lie are well documented, from illegal immigrants to closeted gays and lesbians. As the adage says, cheaters never prosper.

The consensus among several dozen American survey directors at the 2007 Interviewer Falsification Summit was that career interviewer cheaters were extremely rare (AAPOR 2009). I have personally worked with approximately 550 interviewers since 1973 (as an interviewer myself, an interviewer supervisor, and a principal investigator), but I knew just one "calculating cynic." He was a telephone interviewer who did not put enough effort into practicing one-sided conversations. The interviewer one cubicle over ratted him out before he could fake even three interviews. It takes a lot of concerted effort to fake interviews well. In the best SROs interviewers create a proud, self-correcting community dedicated to quality.

It is possible that other interviewer cheaters existed who I never found out. But I doubt it. Persons who have the intelligence to calculate costs versus benefits and who have the diligence to sustain a false identity have probably found employment in higher paying and higher prestige jobs than interviewing.

Interviewers at the other extreme, who try hard but repeatedly make mistakes, the "striving incompetents," have also been rare in my experience, but not as rare. As an example, I once witnessed a new and enthusiastic interviewer mechanically read aloud not only the final question, thank-you, and good-bye as she completed a survey; she also read aloud the interviewer instructions about how to close the interview and go on to the next. Despite re-training and a good attitude, she continued to accumulate such errors and did not make it through the trial employment period. Other such interviewers were down scheduled (i.e., given fewer hours) or not invited to participate in new studies. I suspect many just drifted away because they were not succeeding. It is possible that these people had subtle learning disabilities or attention deficits.

Interviewers who falsify data, in my experience, do so irregularly and without deliberation, and they often admit it. Neither mischief nor mistakes, they misrepresent the truth as a shortcut to exit problematic interviews or fractious respondents. The next section focuses on these interviewers' motivations for their misdeeds.

Interviewers' Real Motivations for Falsifying Data

Having established that few interviewers extend the effort to be a consistent and effective liar, why do some interviewers falsify all or part of the interviews they complete? I classify the reasons as *intrinsic* and *extrinsic* to the job.

Intrinsic items concern the sample, the instrument, the SRO, and respondents. Sample-related items can include inexplicable rules for selecting respondents and the difficulties of locating rare or hard-to-reach populations. At the start of interviews, both interviewers and respondents detest overly long and complicated human subjects explanations. Some topics are inherently less interesting to both interviewers and respondents, and some surveys create significant respondent burden (e.g., retrospective details, overly long, overly sensitive). Particularly annoying for interviewers are poorly designed questions, poorly organized instruments, redundancy across questions, lack of attention to how the instrument appears on the interviewer's computer screen, and inadequate case disposition codes. An inability to voice frustrations or to receive explanations for why an instrument is designed can lead interviewers to skip questions or to commiserate inappropriately with respondents.

SRO-related factors affecting an interviewer's propensity to falsify cover aspects of the interviewer's relationship to the organization. These include inability to get answers to field questions or solutions to computer problems, working on too many different studies, performance pressures, deadline pressures, too many cases, too many difficult cases, unresolved field problems, unclear standards of excellence, and lack of respect and appreciation. Respondent-related factors comprise variations on mean people and pathetic situations. An overdose of these, combined with repeated experiences of futility in the ability to work with such situations, can motivate an interviewer to deviance.

The extrinsic reasons interviewers offer for falsifying all or part of an interview can be grouped into issues associated with place, the family and self, and remuneration. Place-related issues include problematic weather, dangerous neighborhoods, gated communities, and inaccessible apartment buildings. Examples of family-related issues comprise time conflicts and having to work when children are sick or during holiday seasons. Personal issues include feeling unappreciated, fatigued, and disconnected. Of course, interviewers wanting to achieve high performance ratings may also be motivated to innovate. Remuneration-related issues include low pay, timeliness of pay, feeling gypped or hassled about mileage and travel expenses, lack of benefits, lack of job security, and lack of pay security between studies.

These issues intrinsic and extrinsic to the job of interviewing result in differing patterns of falsifying data, described below. Many of the issues could possibly be resolved by expanding interviewers' jobs by drawing on the tacit knowledge they have developed.

Patterns of Falsifying Data: Frequency and Regularity

The reasons interviewers falsify data, above, influence how they do so. Interviewer falsification follows three patterns. These are important to keep this in mind when considering how to monitor and prevent falsification, because one pattern has greater potential damage.

The first pattern of in which interviewers falsify entire interviews is *infrequent and situational*. Their motives may be intrinsic or extrinsic. For example, an interviewer may want to temporarily boost her productivity in an SRO that encourages competition among interviewers to achieve the most telephone interviews in a day. Or a face-to-face interviewer's babysitter needs relief and the last contact of the day is the person she has been unable to reach a dozen times already without success, so she decides to wing it. In such cases, the interviewers are not consistent deviants, and their fake interviews are unlikely to be detected because they falsify too few for patterns to emerge. Fortunately, the damage such falsified interviews inflict to the overall integrity of a survey is minimal, because there are so few cases.

The second way interviewers tend to falsify is *infrequent and systematic*. These interviewers are motivated to avoid only cases they find difficult. Examples include interviewers who cannot abide certain kinds of people (e.g., immigrants with accents), lazy interviewers who pass on the prior refusals to the next interviewer, and fearful interviewers who are deterred by respondents behind barriers, such as gated neighborhoods and guarded apartment buildings. This pattern is problematic because it can create bias for subgroups, such as first-generation immigrants, respondents who initially refused, and people in fortified dwellings. However, such cases should be identifiable with follow-up verification.

The most difficult pattern of interviewer falsification is *frequent and systematic*. These interviewers falsify all or almost all of their completed interviews, creating the potential for meaningful bias in a study's results. This level of interviewer falsification, however, is highly unlikely in SROs who routinely verify completed interviews and conduct post-hoc statistical analysis.

Knowing these theoretical patterns of falsification can be useful for identifying deviant interviewers. However, the extent to which deviant interviewers can be classified into these three categories is unknown and more research is needed.

Equally important to identifying deliberate interviewer misbehavior is identifying interviewers who constantly but accidentally make mistakes. The interviewer whose finger routinely slips off the "4" key to the "1" key also makes frequent and systematic errors that can cause bias in a survey's results, especially if that interviewer has conducted a disproportionate share of the interviews for a survey. Interviewers who systematically err can be just as dangerous to the quality of survey data as those who deliberately cheat.

Preventing and Detecting Interviewer Deviance

SROs prevent interviewers from falsifying data mainly by telling them that procedures are in place to detect cheating. Knowledge of such surveillance is intended to deter interviewer misbehavior. This approach employs various forms of monitoring, re-contact verification, and statistical data analysis to uncover the most egregious interviewer misbehavior. Note that SROs can use the same procedures to correct interviewer mistakes. This section briefly summarizes these methods of detection and some of the issues they present. I conclude with my preferred method of prevention, namely creating an organizational culture in which interviewer deviance is itself unusual.

Monitoring

Foremost, monitoring is intended to verify that an interview occurred; importantly, monitoring can also be used to assess the quality of interviewing. The simplest type of monitoring is management by walking around. There is nothing like a supervisor's continuous presence to deter misbehavior. At the same time, interviewers can ask for immediate help. However, this is possible only in centralized telephone interviewing situations. In addition, an individual observing interviewers dialing, calling, talking, and entering data cannot be everywhere at once and cannot keep objective counts. To enhance human monitoring, SROs employ three basic types of technological monitoring: real-time, manual, and automated.

Real-time monitoring involves using technology to listen in on or watch interviews as they occur. Centralized telephone interviewing facilities offer superior opportunities for interviewer supervision and observation; they typically utilize audio, visual, and screen capture devices to monitor. Field interview settings typically use audio- or video-taping; cameras embedded in laptop computers also allow real-time monitoring of face-to-face interviewing in remote field settings.

The goal of monitoring is to identify unusual, incorrect, or inauthentic behaviors and keystrokes. A problem with both aural and visual real-time monitoring is that misbehaviors occur swiftly and are easy to miss. Another problem is that the listeners/observers are typically physically distant from the interviewing activity and cannot immediately address misbehavior. Both issues undermine the effectiveness of real-time monitoring for quality control, but they can identify areas in which additional interviewer training is needed.

Manual monitoring seeks to identify out-of range patterns. This includes interviewers whose real-time production is out of range, e.g., number of completed interviews per hour, per shift, or per day, number of minutes per completed interview, and dial attempts per completed interview. Interviewers considerably above or below the average are contacted as soon as possible to ascertain they have reasonable explanations for their outliers.

Automated real-time monitoring is probably only effective for detecting problems in a centralized telephone SRO rather than field interview situations. This involves software that charts, for example, number of dial attempts per hour and number of interviews per hour signed into CATI for each interviewer.

Interviewers with out-of-range markers may be faking interviews or making mistakes.

Monitoring raises a number of additional questions, not the least of which is cost, especially in small shops lacking the resources to monitor all interviewers continuously. Who should do it–supervisors or independent outsiders? Who trains the monitors and to what standards? How frequently should each interviewer be monitored and for how long? Should monitors concentrate at the start of studies when interviewers' skills may be rough, at the end when respondents are tough, or throughout data collection? Should monitors concentrate their efforts on new interviewers, those having difficulties, or everyone regardless of their experience? Should interviewers know when monitoring occurs or just know in principle that they will be monitored? Do the principles of informed consent require that respondents know their interview may be monitored? What will be monitored–the data itself (i.e., keystrokes to ensure the accuracy of the link between what respondents say and what interviewers type), the quality of probes and tailoring, and interviewers' contact attempts? When out-of-range behaviors occur, who approaches the interviewer about a potential case of falsification–the monitor, the supervisor, or someone else? How will that person present the case when situations can vary widely? Finally, who monitors the monitors? These remain open questions; answers depend on additional research.

Verifying by Recontact

Currently most SROs verify completed interviews more often than they monitor the data collection process, mainly because monitoring is expensive. Recontacting most often occurs by telephone because it is fast and efficient. When telephone is not possible, face-to-face is recommended (but not always possible). Those doing the verifying are generally senior staff in data collection whose commitment to data integrity is demonstrated.

Typically, 5% to 15% of all cases are verified, both randomly selected and over-selected among interviewers who have experienced disproportionate refusals and those whose honor has been questioned. In addition, some SROs target experienced interviewers, based on the logical but unproven idea that they are most likely to know how to cheat without getting caught. Surveys that are especially long, that contain sensitive questions, and that occur around holidays are likely to have higher percentages of cases verified (Lawrence and Love 2010).

The items typically verified in the callback interview are the study's overall topic, the interview mode, the interview length, incentives received (especially cash payments), and key questions, particularly those associated with lengthy skip logic.

Statistical Data Analysis

The standard statistical quality control methods used in SROs can also be applied to identify out-of-range patterns that may indicate interviewer error or subversion. Routine reports using paradata, such as time stamps, productivity indicators, refusal rates, and case disposition codes, can include daily Schwart charts and weekly Pareto analyses, produced either manually or automated depending on the SRO's capabilities. Such tracking mechanisms for daily or weekly production can identify interviewers who complete interviews in too few or too many minutes, who complete questions in too few or too many seconds, whose telephone interview completions per dial attempt are unusually efficient, whose mileage per completed face-to-face interview is relatively low, and other creative paradata approaches (Murphy, Eyerman, McCue, and Hottinger 2005).

Post-hoc statistical analysis of survey data is also used to flag suspicious seeming answers to questions. These analyses include identifying outliers, testing for counter-intuitive responses, examining patterns of variance within and between respondents, investigating items controlling skip logic on significant blocks of questions, and locating otherwise suspiciously patterned answers (Blasius and Thiessen 2012, Krejsa, Davis, and Hill 2009, Porras and English 2004). The value of testing for outliers is limited, because there are too many reasons for respondents' answers to truly be one or two standard deviations outside the mean and thus have no predictive relationship with problematic interviewers. Examining unlikely or rare combinations of responses, e.g., elderly people with no health problems and heavy smokers with vigorous daily exercise, is inefficient and uneconomical for locating false interviews because devious interviewers know to avoid these unlikely combinations and because testing requires advance knowledge of each survey's contents. The lack of variance approach seems most promising, as described in several chapters of this volume (also Schäfer, Schräpler, Müller, and Wagner 2005).

Because post-hoc data analysis takes place, by definition, after a study is completed, it is not useful at all in solving an immediate problem situation with an interviewer. By the time the analysis occurs, the problematic interviewers may already have departed their SRO employer. GfK Software Alberta 2.0, described elsewhere in this volume, appears to be a promising innovation for applying statistical data mining techniques in real time (Hülser 2013, this volume).

Ultimately, could the time and effort invested in identifying falsification be better spent in creating an organizational environment that prevents cheating and self corrects potential cheaters?

Organizational Culture

Monitoring, verification, and statistical data analysis to prevent and detect data falsification focus SROs' resources (often scarce resources) on identifying interviewers' bad behaviors rather than on recognizing and rewarding their good behaviors. Yet learning theory indicates that humans learn most effectively when they receive positive reinforcement (Skinner 1976). It is possible to use the evidence gleaned from the same procedures to identify excellent interviewers and create interviewer training opportunities, but the extent to which SROs do so routinely is unknown. It is known that moving away from a corrective orientation requires a broad shift in management focus and organizational culture.

At the root of creating an organizational culture that minimizes data falsification is the recognition of the value of interviewers' work to the SRO, including an open acknowledgment of the significance of the fact that SROs connect to respondents primarily through interviewers. SROs must also demonstrate respect of the fact that interviewers are highly skilled on an array of projects with a diverse assortment of respondents. The best SROs invest in understanding interviewers' work as they experience it (Converse and Schuman 1974) and use that to create a positive employment experience for interviewers.

Even though interviewers are the backbone of SROs, they remain invisible to most principle investigators and SRO upper management. Interviewers deliver the carefully crafted survey questions that operationalize investigators' concepts; they employ considerable tacit knowledge with swift perspicacity to identify respondents' out-of-bounds answers and guide them back on target with probes. Yet few investigators bother to meet their survey's interviewers, explain their studies, or thank them. The best evidence for how SROs generally undervalue interviewers is in the fact that most work on a part-time, temporary basis, at odd hours, and without benefits or job security and most earn minimum wage. Many SROs can terminate interviewers' employment without cause and without advance notice.

Thus, one important part of preventing interviewer deviance is to demonstrate their value to the organization. Of course, paying well helps, but not all SROs can pay living wages, offer benefits, or provide continuous employment. In such a situation, the SRO can use non-financial motivators, such as awards, t-shirts, and a genuine thank you to recognize a job well done. Learning theory demonstrates that intermittent rewards tend to inspire the desired behavior better than continuous or predictable incentives. Even if the SRO is demonstrably poor in its ability to reward interviewers in a manner they deserve, it helps to offer even the most modest hourly increases in wages in predictable fixed intervals, e.g., a raise of $.05 for every 1,000 hours of interviewing.

Similarly, an interviewer manual that documents standards of excellence is essential. Clearly stated desirable behavior in all aspects of the interviewing job clarifies the rationale for monitoring. Moreover, defining multiple important facets of the job (including, for example, voice inflection and pace of speech) creates a new rationale for monitoring; i.e., it is not focused solely on falsification, but also on providing feedback on various facets of job performance that will help identify and minimize interviewer error

Involving interviewers in more survey research activities than simply interviewing will increase their hours and pay and, at the same time, help to ameliorate some of the intrinsic and extrinsic aspects of the job that may propel some interviewers to behave deceitfully. I suggest, for example, involving interviewers with the rigorous pretesting of instruments and of instruments as programmed into software. Who is better to identify problems with the instrument than those who use it daily? Even if their suggestions on how to revise or reorder questions go unheeded, having a voice in the process recognizes the tacit knowledge they have earned in the field. Interviewers can also offer assistance in devising sample selection rules based upon their extensive experience implementing those rules. Ultimately, the sample selection rules may still end up complex, but the interviewers will have a vested interest in understanding why and in making them work.

Including interviewers in the very necessary but non-interviewing aspects of the survey process lessens the social distance between research scientists and the interviewers. Another step in the same direction is to require survey clients (i.e., agency representatives and principal investigators) to attend at least part of interviewer training for their survey. Consistently, in my experience, clients leave such sessions confident that data collection for their project will go well.

These means of acknowledging that interviewers are professionals in their field reduces status-related alienation that motivates some individuals to mischief. Even though they may still chafe at pressures and unresolved field problems, there will be fewer intrinsic issues to cumulate and promote deviant behavior.

Conclusion

It is important to reiterate here that interviewer falsification is rare. Even where it occurs it may not harm data as much as systematic interviewer error. This chapter has attempted to place such issues associated with interviewers in the broader context of total survey error, recognizing that many people are involved in the life cycle of interviewer-administered surveys and each can contribute to survey error. Most of this non-sampling error remains difficult to quantify, but before investing intense efforts into uncovering interviewer-related error structures and rates, it is important to assess the relative importance of various error sources.

Our ultimate concern is overall survey quality and creating an organizational culture in SROs in which data falsification by interviewers is unknown. Toward that goal, survey researchers must keep in mind that interviewers are the link between our ideas and our data. Without properly trained, remunerated, and respected interviewers, we cannot trust the data we use to test hypotheses and evaluate public policies. Post-hoc data mining to identify patterns of falsification and locate potential interview fakers are too late; the data are already contaminated and the perpetrator may be already looking for an excuse to create mischief at another SRO.

I have attempted to identify characteristics of SROs that maintain a corps of productive interviewers, that maximize interviewers' motivation to adhere to structured conversational routines, and that minimize interviewers' potential inclination to deviate from those routines. The most salient characteristic of such SROs is that interviewers' work is respected for what it is, interviewers are not segregated from scientists, and interviewers' skills are used between episodes of data collection on other tasks to maintain stable employment. Such features of SROs are essential if survey researchers are to maintain the public's trust.

Bibliography

AAPOR (American Association for Public Opinion Research): Interviewer Falsification–Practices And Policies, http://www.aapor.org/Content/aapor/AdvocacyandInitiatives/StandardsandEthics/InterviewerFalsificationPracticesandPolicies/default.htm, 2009. (Accessed 5/31/2012)

Andersen, Ronald, Kasper, Judith, Frankel, Martin R. and Associates: Total Survey Error, San Francisco, Jossey-Bass, 1979.

Biemer, Paul P.: Total survey error: design, implementation, and evaluation, Public Opinion Quarterly, 74, 817-848, 2010.

Blasius, Jörg: Detecting (partly) faked interviews, paper presented at Workshop: Interviewers' Deviant Behaviour–Reasons, Detection, Prevention, Castle Rauischholzhausen – Conference Centre of the Justus-Liebig University of Giessen, 2011.

Blasius, Jörg and Thiessen, Victor: Assessing the Quality of Survey Data, London, Sage, 2012.

Brüderl, Josef, Huyer-May, Bernadette, and Schmiedeberg, Claudia: Interviewer Behavior and the Quality of Social Network Data, In: Winker, Peter, Menold, Natalja, and Porst, Rolf (eds.): Interviewers' Deviations in Surveys – Impact, Reasons, Detection and Prevention. Frankfurt/Main: Peter Lang, 2013. pp. 147-160.

Converse, Jean M. and Schuman, Howard: Conversations at Random: Survey Research As Interviewers See It, New York, Wiley, 1974.

Crespi, Leo P.: The cheater problem in polling, Public Opinion Quarterly, 9, 431-445, 1945.

Dillman, Don A., Smyth, Jolene D. and Christian, Leah Melani: Internet, Mail, and Mixed-Mode Surveys: The Tailored Design Method, New York, Wiley, 2009.

Eisenhauer, Joseph G.: The shadow price of morality, Eastern Economic Journal, 32, 437-456, 2006.

Gwartney, Patricia A.: The Telephone Interviewer's Handbook: How to Conduct Standardized Conversations, San Francisco, Jossey-Bass/Wiley, 2007.

Hülser, Oliver: Automatic interview control in market research studies, In: Winker, Peter, Menold, Natalja, and Porst, Rolf (eds.): Interviewers' Deviations in Surveys – Impact, Reasons, Detection and Prevention. Peter Lang, 2013. pp. 103-116.

Krejsa, Elizabeth A., Davis, Mary C., and Hill, Joan M.: Evaluation of the quality assurance falsification interview used in the Census 2000 Dress Rehearsal, paper presented at the Joint Statistical Meetings, Section on Survey Research Methods. http://www.amstat.org/sections/srms/proceedings/papers/1999_110.pdf, 2009. (Accessed 5/31/2012)

Lawrence, Chrishelle and Love, Elizabeth: Characteristics of falsified interviews, paper presented at the Joint Statistical Meetings, Section on Survey Research Methods. http://www.amstat.org/sections/srms/proceedings/y2010/Files/308726_61475.pdf, 2010. (Accessed 5/31/2012)

Maynard, Douglas W., Houtkoop-Steenstra, Hanneke, Schaeffer, Nora Cate and van der Zouwen, Johannes (eds.): Standardization and Tacit Knowledge: Interaction and Practice in the Survey Interview, New York, Wiley Interscience, 2002.

Maynard, Douglas W. and Schaeffer, Nora Cate: Standardization-in-interaction: The survey interview, In: Drew, Paul, Raymond, Geoffrey and Weinberg, Darin (eds.): Talk and Interaction in Social Research Methods (9-27), London, Sage, 2006.

Murphy, Joe, Baxter, Rodney, Eyerman, Joe, Cunningham, David and Kennet, Joel: A system for detecting interviewer falsification, paper presented at the American Association for Public Opinion Research 59th Annual Conference, http://www.rti.org/pubs/paper-detectingintrvfalsifi.pdf, 2004. (Accessed 5/31/2012).

Murphy, Joe, Eyerman, Joe, McCue, Colleen, Hottinger, Christy and Kennet, Joel: Interviewer falsification detection using data mining, Proceedings of Statistics Canada Symposium 2005: Methodological Challenges for Future Information Needs, http://www.statcan.gc.ca/pub/11-522-x/2005001/4199022-eng.pdf, 2005. (Accessed 5/31/2012)

Porras, Javier and English, Ned: Data-driven approaches to identifying interviewer data falsification: The case of health surveys, paper presented at the Joint Statistical Meetings, Section on Survey Research Methods. http://www.amstat.org/sections/srms/proceedings /y2004/files/Jsm2004-000879.pdf, 2004. (Accessed 5/31/2012)

Schäfer, Christin, Schräpler, Jörg-Peter, Müller, Klaus-Robert and Wagner, Gert G.: Automatic identification of faked and fraudulent interviews in the German SOEP, Schmollers Jahrbuch, 125, 183–193, 2005.

Skinner, B. F.: About Behaviorism, New York, Vintage Books, 1976.

Tourangeau, Roger, Rips, Lance J. and Rasinski, Kenneth A.: The Psychology of Survey Response, New York, Cambridge University Press, 2000.
Weisberg, Herbert F.: The Total Survey Error Approach: A Guide to the New Science of Survey Research, Chicago, University of Chicago Press, 2005.

Extreme Response Style and Faking: Two Sides of the Same Coin?

Matthias Ziegler and Christoph J. Kemper

Abstract

Using Mixed Rasch Models, several studies identified two distinct classes present in personality test data collected under lowstakes (i.e. the test person has nothing to gain or lose based on the test results) and anonymous conditions. Based on threshold diagrams the classes were interpreted as two response styles – extreme response style and midpoint responding. Within psychological research concerned with faking, there are some studies in which Mixed Rasch Models were also applied. However, in those studies data were derived from high stakes (i.e. the test person can gain or lose something based on the test results, e.g., a job application situation) and non-anonymous conditions. Again, two response bias classes emerged. This time they were interpreted as representing slight and extreme faking. Whereas slight fakers deviate only little from their actual honest answer in a personality questionnaire, extreme fakers deviate strongly. It is unclear, though, in how far these findings actually represent distinct response biases. Nevertheless, Mixed Rasch Models might offer the chance to correct survey data for these response biases. The present chapter shortly summarizes prior findings and then reports a study in which both conditions, i.e. low and high stakes settings, were realized within one sample, allowing to investigate the relationship between the different biases found in different conditions. The findings suggest that the previously reported response styles for different high and low stakes conditions might be less distinct than assumed. This raises the question whether the Mixed Rasch Model can really capture faking. Practical implications are considered.

Introduction

Questionnaires are an efficient and widely applied source of research data in the social sciences. Behavior, attitudes, or personality are usually measured by having respondents indicate their level of agreement to a series of statements on a multi-point rating scale. However, variance due to a substantive construct is not the only source of systematic variance in these measurements. Observed responses and scores may also contain variance due to response bias. The present book deals with these error sources mainly from a perspective in which an interviewer is thought to have manipulated the data. While this is extremely important, it should also be kept in mind that the interviewee might distort answers to the questions. This perspective is taken in the present chapter. From the perspective of the interviewee different response biases can be distinguished. Paulhus (1991) refers to response bias as a systematic tendency to respond to questionnaire items on some basis other than the specific content. This systematic tendency might be a temporary reaction to a situational demand (response set) or a stable and consistent behavior (response style). For example, in high stakes situations, i.e., situations in which the test taker can gain or lose something based on the test results, such as job applications (Zickar, Gibby, & Robie, 2004) persons may deliberately distort their answers in order to create a positive impression. This response bias is referred to as faking (MacCann, Ziegler, & Roberts, 2011). A more general term is social desirability. Yet, social desirability can also include unconscious distortions like self-enhancement. Therefore, we use the term faking to clearly indicate that a willful distortion is meant. Two distinct classes of fakers were identified in previous studies – slight fakers (SF) and extreme fakers (EF) (Robie, Brown, & Beaty, 2007; Ziegler, 2011). Slight fakers are described as people who distort their responses only little. Extreme fakers, however, strongly deviate from what would actually be their honest response. In low stakes situations, i.e., there is nothing to gain or lose for the test taker, response biases may occur due to stable and consistent preferences of persons to answer rating scale items in a specific way. For example, some persons prefer extreme response categories of a rating scale while others prefer moderate categories to describe themselves (extreme responding, ERS; midpoint responding, MPR). Numerous studies have shown that these biases occur when persons respond to rating scale items and that they may contaminate the measurement of substantive constructs (Austin, Deary, & Egan, 2006; Gollwitzer, Eid, & Jürgensen, 2005; Meiser & Machunsky, 2008; Rost, Carstensen, & Von Davier, 1997). The existence of SF and EF as well as ERS and MPR raises at least two interesting questions for the survey context. On the one hand, if such individual response styles could be modeled, survey data could be corrected for such distortions. On the other hand, especially ERS is also of interest within survey research concerned with identifying interview protocols faked by the interviewer (Bredl, Winker & Kötschau, 2012). Thus, if the same response bias can be found for the interviewee, more care is needed when using ERS to identify dishonest interviewers.

Thus, response biases threaten the data quality of surveys and the validity of conclusions based on survey data. Moreover, such response biases can be found on the level of the interviewee as well as on the level of the interviewer. In the research presented in this article, we focus on two interviewee response biases that have been identified applying Mixed Rasch Models (MRM), i.e. ERS/MPR and SF/EF in previous research. Our research questions are (1) whether people responding to a questionnaire with rating scale items in a low and a high stake situation can be classified as belonging to a specific response bias class, and (2) how these classifications of ERS/MPR and SF/EF are related. The findings bear practical implications for survey research. If it is possible to identify such response biases, correction methods could be applied to improve data quality in surveys.

On the following pages we will first shortly summarize findings with regard to ERS and MPR. Afterwards we will introduce the basic ideas of the Mixed Rasch Model followed by a short overview of psychological faking research applying this method. Finally, we will present results from an empirical study where both response styles, i.e. ERS/MPR and SF/EF, were studied on the level of the test taker. The chapter closes with a discussion of these results as well as some practical implications.

Extreme Response Style and Midpoint Responding

ERS refers to the tendency to disproportionately favor the endpoints or "extreme" categories of rating scales, irrespective of particular item content (Naemi, Beal, & Payne, 2009). In contrast, MPR refers to the tendency to prefer middle or neutral categories of rating scales. Both response styles are an issue of concern for researchers using questionnaires as a source of research data. Distinct groups of persons who prefer to describe themselves with either extreme or moderate ratings were found for diverse constructs, e.g. the Big Five personality dimensions (Austin et al., 2006; Hinz, Brahler, Geyer, & Korner, 2003; Rost, Carstensen, & von Davier, 1999b), personal need for structure (Meiser & Machunsky, 2008), anger expression (Gollwitzer et al., 2005), anxiety sensitivity (Kemper, 2010), and in organizational surveys (Eid & Zickar, 2007). These results suggest that ERS and MPR might affect the measurement of constructs in general, especially when multi-point rating scales are used. ERS and MPR may introduce additional construct-irrelevant variance to the measurement leading to biases on the individual and on the group level. On the individual level, a person's tendency to prefer extreme categories may result in a higher score and thus in a higher level of the intended-to-be-measured trait (Bolt & Johnson, 2009). On the group level, another bias might occur. When items function differently in subgroups of the population, observed effects can be a result of group specific response styles. In this case, mean differences between groups cannot be validly compared (Bolt & Johnson, 2009; Eid & Rauber, 2000). As underlying dispositional factors of ERS/MPR Naemi et al. (2009) propose so-called epistemic constructs (Kruglanski, 1989), e.g. judgment complexity, rigidity, intolerance of ambiguity, and dogmatism. Such dispositions describe the manner in which people process and respond to information. Besides possible dispositional antecedents of these response styles, researchers have explored their relations to various individual difference variables. Substantial correlations were found between ERS/MPR and sex, age, cognitive ability, education, ethnicity, and personality, especially extraversion and anxiety (Austin et al., 2006; Berg & Collier, 1953; Eid & Rauber, 2000; Greenleaf, 1992; Meisenberg & Williams, 2008; Naemi et al., 2009). However, results are considerably less consistent and conclusive regarding the nature and direction of the observed relations.

The present study will not focus on possible relations between ERS/MPR and personality characteristics. Instead, the relationship with SF/EF will be investigated to determine the uniqueness of these supposedly distinct response biases.

The Mixed Rasch Model

The most sophisticated method to identify ERS and MPR is the so-called Mixed Rasch Model (MRM; Rost et al., 1997). The MRM combines a Latent Class Analysis (LCA) with a Rasch analysis. In the present analyses a LCA was combined with an ordinal Rasch model. The basic equation is:

$$p(X_{vi} = x) = \sum_{g=1}^{G} \pi_g \cdot \frac{e^{(x \cdot \theta_{vg} - \sigma_{ixg})}}{\sum_{s=0}^{m} e^{(m \cdot \theta_{vg} - \sigma_{img})}}$$

$p(X_{vi} = x)$ = Probability of person v to chose category x for item i

$\sum_{g=1}^{G}$ = Sum from class g = 1 to class G

π_g = Class size parameter

σ_{ixg} = Sum of all threshold parameters of item i up to chosen category x in class g

θ_{vg} = Person parameter for person v in class g

σ_{img} = Sum of all threshold parameters of item i in class g

m = Number of thresholds

s = threshold

Consequently, in MRM different qualitative classes can be assumed in which person parameters may vary quantitatively. Results of the MRM analysis are indicative of whether there are subgroups in the sample data for which the items are differentially difficult to answer or subgroups that exhibit different response behaviour. To determine the number of classes representing the data best, information criteria, e.g. the Consistent Akaikes Information Criterion (CAIC) are usually compared (Rost et al., 1997; Zickar et al., 2004). Thus, this combination of a Latent Class Analysis and a Rasch model allows for qualitatively distinct classes in which people can differ quantitatively on the trait being measured. As could be shown, under certain conditions, the results of an MRM analysis can be interpreted as evidence of a response bias measured along with a trait.

Using MRM, two types of information can be obtained for each individual. Each person is assigned to one class and within each class a person parameter representing the individual's trait level is estimated. Person parameters theoretically range from minus to plus infinity but scores between minus and plus 3 are most common. A score of zero indicates an average trait level. The higher the score, the higher the trait level. On the item level, threshold parameters are estimated for each item. The item thresholds represent the boundary values between neighboring rating categories of an item. Thresholds are scaled on the same dimension as the person parameters and represent the level a person parameter must exceed to achieve a higher probability for the person to pass from one rating category to the next, e.g. from "very likely" to "extremely likely". Because the classes differ qualitatively, threshold values cannot be compared between the classes. However, it is possible to find out whether the classes found in the data emerge due to differences in traits or differences in response bias by inspecting the item threshold diagrams. If the thresholds run parallel in both classes, this indicates a response bias. Most studies that found ERS/MPR could show that the estimated threshold parameters run parallel in these two classes. This is interpreted as indicating two classes that only differ in their response bias but not the trait actually measured. This method was also used in the research presented here.

Extreme and Slight Faking

MacCann et al. (2011) define faking as: "... a deliberate set of behaviors motivated by a desire to present a deceptive impression to the world. Like most other behavior, faking is caused by an interaction between person and situation characteristics. (p.311)". Faking is further seen as a conscious effort to manipulate responses to self-descriptive questionnaire items (Ziegler, 2011; Ziegler, MacCann, & Roberts, 2011). The ultimate goal of faking is to create a specific impression depending on the demands faced by a person in everyday situations. For example, the deliberate promotion of competence, fearlessness, and physical prowess is often seen in job applicants or males trying to impress a dating partner. Another example is the deliberate minimization of faults involving excuse-making and damage control. This behavior might be seen in religious settings or legal defendants trying to avoid punishment (Paulhus, 2002). According to Paulhus, the former is termed agency management and the latter communion management. Both forms of impression management or faking are considered a problem for researchers working with questionnaire data as they may introduce a systematic error to the measurement of substantive constructs (Heggestad, George, & Reeve, 2006). When faking questionnaires, persons may distort item scores and aggregated scale scores slightly or extremely (slight vs. extreme fakers) impacting on the criterion (Ziegler, Danay, Schölmerich, & Bühner, 2010) and construct validity (Ziegler & Bühner, 2009) of the measurement. Intensive research focused on the underlying factors and processes of faking behavior. Concerning situation attributes, a psychological demand is necessary motivating and pressuring the person to create a certain impression, e.g. looking more favorable or impressive than one actually is (faking good) or looking more problematic or psychologically troubled than one actually is (faking bad) (Ellingson, 2011; Mueller-Hanson, Heggestad, & Thornton, 2006). Concerning person attributes, faking seems positively related to self-efficacy of positive self-presentation (Pauls & Crost, 2005; Ziegler, 2007), self-monitoring, and cognitive ability, and negatively related to integrity (McFarland & Ryan, 2000). As far as Big Five traits are concerned, results are less consistent. However, all Big Five traits were found to be related to faking, i.e. affected by it (Birkeland, Manson, Kisamore, Brannick, & Smith, 2006; Grubb & McDaniel, 2007).

As was the case with ERS and MPR MRMs are often used to identify slight and extreme faking in questionnaire data. In both cases, resulting classes are interpreted based on the fact that the item thresholds for the found classes run parallel, indicating the same trait is measured but with the impact of a response bias. For the data acquired under low stakes conditions the classes are called ERS and MPR. For data derived from high stakes settings SF and EF. However, the differences between the respective classes oftentimes look pretty comparable. Despite the fact that the methods used and the way results are interpreted are the same, the relationship between ERS/MPR and SF/EF is unknown. If the two response biases were indeed distinct, it would be possible to distinguish between a mere response style and actual faking. However, if the biases were related, MRMs could still be used with survey data to correct the scores if either one is thought to be likely to have occurred.

Method

Participants and Procedure

326 undergraduate psychology students of a German University participated in the study. Due to technical problems during the administration of the measures (see below) used, 14 data sets were lost. The final sample consisted of N = 312 participants between 21 and 53 years of age (M = 25, SD = 5.5). 79% were females. Participants were invited to a lab on campus and randomly assigned to either an experimental group (n_E = 157, 121 females) or a control group (n_C = 155, 126 females). After providing informed consent, participants were seated in front of a computer and received a briefing. A test battery containing several tests was administered on the computer. Part of this battery was the Revised NEO Personality Inventory (NEO-PI-R; Ostendorf & Angleitner, 2004). Several other scales which are not the focus of the present research were also administered (e.g., the Intelligence Structure Test 2000-R (IST 2000-R; Amthauer, Brocke, Liepmann, & Beauducel, 2001)). For the NEO-PI-R, participants received different instructions. In the control group, participants were instructed to fill out the NEO-PI-R honestly on the first measurement occasion as well as on the second. In the experimental group, participants were instructed to respond to the personality test honestly on the first occasion (low stakes situation) and fake-good on the second occasion (simulated high stakes situation). Thus, data from the low stake situation can be used to model ERS and MPR. Data from the simulated high stakes situation can be used to model SF and EF. The faking instruction used in the simulated high stakes condition contained a realistic scenario for all participants with a warning to discourage obvious faking as proposed by Rogers (1997). As participants were university students, we adapted the instruction to this context: "Universities have to select their students. To achieve a good selection, a number of instruments like the following are usually used. Please imagine that you are participating in a student selection procedure. Of course, it is your goal to get an admission as a psychology student. Therefore, you have to fill out the following questionnaire in a way that assures your admission. However, you have to be careful since a test expert will check the results for obvious faking and you do not want to be identified." After completing the test battery, participants received course credit for their participation in the study and were debriefed.

Measures

Personality was assessed with a German adaptation (Ostendorf&Angleitner, 2004) of the NEO-PI-R (Costa & McCrae, 1992). The NEO-PI-R allows for a comprehensive assessment of personality based on the Five-Factor Model (FFM). Containing 240 items, the NEO-PI-R measures the Big Five personality dimensions of Neuroticism, Extraversion, Openness, Agreeableness, and Conscientiousness with 48 items per dimension. Each dimension consists of several facets, e.g. Neuroticism consists of Anxiety (N1), Hostility (N2), Depression (N3), Self-conscientiousness (N4), Impulsiveness (N5), and Vulnerability (N6). Each of these facets is measured by 8 items. Items contain statements and a five-point rating scale for respondents to indicate their endorsement ranging from "strongly disagree" to "strongly agree". The NEO-PI-R is a comprehensive, reliable and valid measure of the Big Five personality dimensions and its facets (for details on the psychometric quality see Costa & McCrae, 1992; Ostendorf & Angleitner, 2004). Internal consistencies here were comparable to those reported in the German manual and ranged from $\alpha = .46$ to $\alpha = .88$ for the personality facets.

Statistical Analysis

Data was analyzed with SPSS 20. Furthermore, the computer program WINMIRA (von Davier, 2001) was used for the MRM and the LCA. To avoid response bias associated with the middle category of the five-point rating scale other than MPR/ERS or SF/EF, e.g., its correlation with item clarity (Kulas & Stachowski, 2009), the scale was transformed to a four-point scale for the MRM analyses as proposed by Rost, Carstensen and von Davier (1999a) as well as Austin et al. (2006). This was accomplished by collapsing category 2 with 3. Models with one to four classes were estimated for the data and the best fitting solution selected based on information criteria – Akaike's Information Criterion (AIC), Schwartz Information Criterion (BIC) and Bozdogan's Consistent AIC (CAIC) (for details see Bozdogan, (1987); Read & Cressie, (1988)). Threshold parameter profiles were compared between classes to specify the nature of the found classes. These analyses were conducted for all participants using the data from measurement occasion 1. Here no specific instruction other than to respond honest to NEO-PI-R items was given. Thus, we were looking for two class solutions resembling ERS and MPR. These analyses were run for all 30 facets of the NEO-PI-R separately. As a consequence, each person was classified as either applying ERS or MPR for those facets where these biases could be found. We then used a Latent Class Analysis with these classifications to determine whether people consistently used a response style across facets or switched between styles. Using only data from the experimental group (fake-good condition), the same analyses were conducted for data from occasion 2. This time, the two class solution was supposed to represent SF and EF. Thus, after the analyses two classifications were made for each participant from the experimental group. Each participant was classified as either ERS or MPR based on the data from occasion 1 (low stake situation, i.e. honest responses can be assumed) and as SF or EF based on data from occasion 2 (simulated high stakes situation, i.e. faked responses can be assumed). These classifications of ERS/MPR and SF/EF were then compared using a McNemar test.

Items yielding estimation problems in the MRM analyses or causing class solutions based on content (i.e. the thresholds did not run parallel in the classes) were excluded from the analyses. If a facet had less than five items remaining, the facet was also excluded from further analyses because scale length is an important factor in the accurate identification of classes with WINMIRA (Zickar & Burnfield, 2003).

Results

Using data from occasion 1 ERS/MPR could be shown to occur in 12 facets. For eight out of the 30 facets, 2-class solutions could be found without any further changes. However, for some facets estimation problems occurred for some items which were then dropped from the analyses. As a consequence four facets were no longer analyzed because less than 5 items remained. Having eliminated items with estimation problems 2-class solutions that could be interpreted as ERS/MPR emerged for four more facets. The remaining facets either had 1-class solutions as best fitting result or were not clearly interpretable as ERS/MPR. The 1-class solutions could be due to the small number of items and persons (for a detailed description see Gerber-Braun, 2010). Thus, classifications in 12 facets were the basis for the LCA. LCA results can be found in Table 13.1.

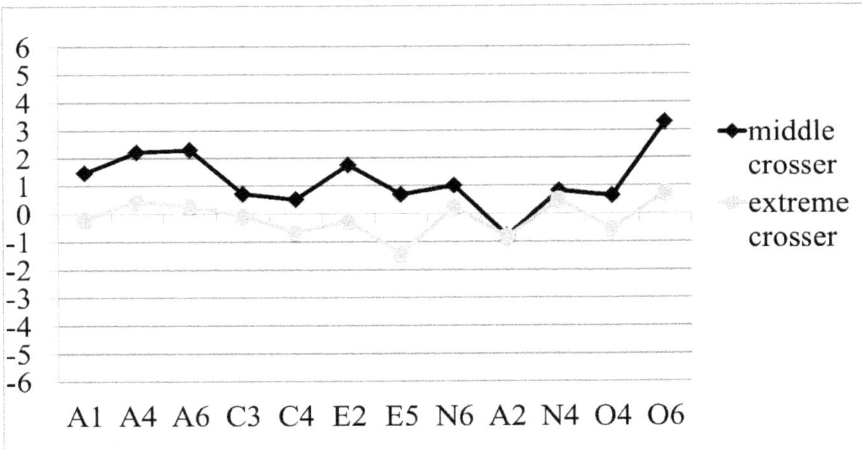

Fig. 13.1: Item locations for the two-class solution of the LCA for the 12 facet classifications; A1 = Trust; E2 = Gregariousness; A2 = Straightforwardness; C3 = Dutifulness; N4 = Self-conscientiousness; O4 = Actions; A4 = Compliance; C4 = Achievement striving; E5 = Excitement seeking; N6 = Vulnerability; O6 = Values; A6 = Tender-mindedness

Table 13.1: Latent Class Analysis with 12 facets of the NEO-PI-R for Occasion 1

Number of classes	*class nr.*	*class size*	*sum score*	*Information Criteria*		
				AIC	*BIC*	*CAIC*
1	1	1	4.27	4537.13	4582.04	4594.04
2	1	.67	3.21	4426.02	**4519.60**	**4544.60**
	2	.33	6.40			
3	1	.40	3.47	**4424.70**	4566.93	4604.93
	2	.31	6.54			
	3	.29	2.95			
4	1	.35	3.39	4425.18	4616.07	4667.07
	2	.23	5.23			
	3	.23	2.32			
	4	.19	6.99			

Notes: sumscore = sum of the means of the facets included in the LCA; bold class solutions indicate the best solution according to the particular criteria

Table 13.1 shows that the two-class solution fitted the data best according to the commonly chosen CAIC. Moreover, Figure 13.1 shows that the item locations for the two classes run almost parallel. This is usually seen as representing a response style. Thus, the analyses support a two class solution with about one third of participants being classified as applying ERS.

Before conducting the analyses with the data from occasion 2 (simulated high stakes), ANCOVAs for all facets were conducted to identify facets with substantial differences in personality scores between the experimental and control group. Data from Occasion 2 served as dependent variables and data from Occasion 1 as covariate. Experimental vs. control group was used as independent variable. Only those facets with significant and substantial differences between both groups after controlling for baseline differences were considered to be affected by faking and included in the subsequent MRM analyses. Only two of 30 facets were not faked (Openness to aesthetics and Openness to values, $p > .05$, partial $eta^2 < .013$). Thus, there were substantial and significant differences between the experimental and the control group in 28 of the 30 facets indicating faking. After the following analyses mirroring the procedure described above, thirteen facets yielded two class solutions consistent with an interpretation of classes as SF and EF. Classifications from these thirteen facets were used for the LCA. Results can be found in Table 13.2.

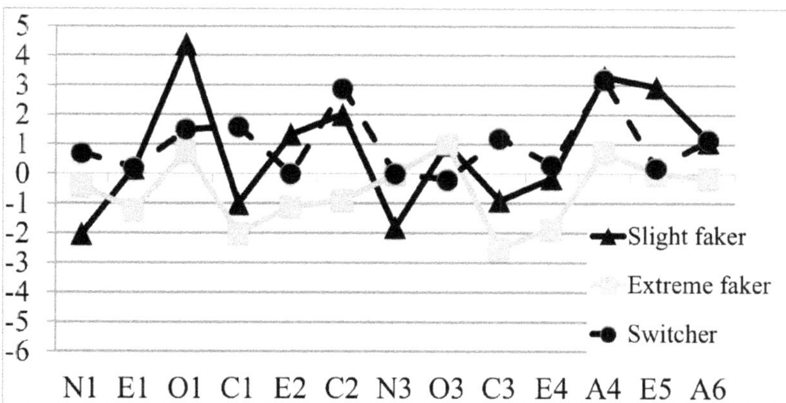

Fig. 13.2: Item locations for the switchers, slight fakers and extreme fakers of the 3-class solution of 13 facets in the LCA; N1 = Anxiety; E1 = Warmth; O1 = Fantasy; A1 = Trust; C1 = Competence; E2 = Gregariousness; C2 = Order; N3 = Depression; O3 = Feelings; C3 = Dutifulness; E4 = Activity; A4 = Compliance; C4 = Achievement striving; N5 = Impulsiveness; E5 = Excitement seeking; O5 = Ideas; A6 = Tender-mindedness.

Table 13.2: Latent class analysis with 13 facets of the NEO-PI-R for Occasion 2

Number of classes	*class nr.*	*class size*	*sum score*	*Information Criteria*		
				AIC	*BIC*	*CAIC*
1	1	1.00	5.25	5028.01	5076.67	5089.67
2	1	0.67	4.30	4839.51	4940.57	4967.57
	2	0.33	7.18			
3	**1**	**0.17**	**4.17**	4754.87	**4908.34**	**4949.34**
	2	**0.53**	**5.13**			
	3	**0.30**	**7.88**			
4	1	0.37	4.36	4722.98	4928.84	4983.84
	2	0.23	5.45			
	3	0.23	4.31			
	4	0.16	8.31			
5	1	0.25	5.49	**4707.36**	4965.63	5034.63
	2	0.24	4.47			
	3	0.21	5.83			
	4	0.20	3.27			
	5	0.12	8.69			

Notes: sum score = sum of the means of the facets included in the LCA; bold class solutions indicate the best solution according to the particular criteria

The findings support a 3-class solution. As can be seen in Figure 13.2, two classes again have almost parallel item locations indicative of a response style. One class however has item locations that are not parallel. The most likely explanation is that this class used a response style for some of the facets (here parallel item locations occur) but did not use a response style for other facets. In other words, this class did not fake on all of the facets. Based on MRM classifications, 53% of participants were consistently classified as SF and 30% as EF by the Latent Class Analysis. 17% of the participants switched from responding honestly for some facet to responding dishonestly for others. This is most likely due to differences in interpreting the meaning of the facet with regard to the utility of the trait for the imagined student selection procedure (Ziegler, 2011).

Summing up, the first part of the analyses, participants could be classified consistently as using ERS or MPR for most facets based on data from Occasion 1. Classification as SF or EF with data from Occasion 2 (faking-good) was consistent for most participants but not for all. These switchers were excluded from the following comparison between ERS/MPR and SF/EF as our main interest was the statistical relation between these response biases. Thus, n = 129 participants remained for the McNemar-Test. Table 13.3 contains a cross-classified table showing the change of the classification from Occasion 1 to Occasion 2.

Table 13.3: Cross-classified table for the change of classification from Occasion 1 to Occasion 2

	Faking (Occasion 2)		*Σ*
Response Style (Occasion 1)	*SF*	*EF*	
MPR	64	24	88
ERS	18	23	41
Σ	82	47	129

The McNemar-Test yielded a small and nonsignificant result ($\chi 2 = .60$, $p = .44$, Cohen's $w = .07$). There was no substantial change in classification between the measurement occasions. Thus, there is no difference between the two response biases ERS/MPR and SF/EF that could not be explained by chance.

Discussion

The present analyses were conducted to shed light on the relationship between two well investigated response biases. ERS and MPR have repeatedly been found in research data collected in low stakes settings using Mixed Rasch Models. With the same technique, slight and extreme faking (SF/EF) have been found in data derived from high stake settings. Despite the overlap in the interpretation of the underlying statistical analyses, the relationship between the two response biases is unknown.

The findings presented here suggest that the supposedly distinct response biases occurring under low stake (ERS/MPR) and high stake conditions (SF/EF) are not that distinct after all. In fact, the results strongly suggest that people use the same response style regardless of the demand they face in specific situations, e.g. when responding to a survey questionnaire in a low stake situation (computer assisted web interview) or a high stake situation (face-to-face interview with family members or peers present). An exception in our study was the small class of participants who switched from responding honestly to dishonestly only for some facets.

These results bear little hope to use the MRM to correct for faking in survey data or even to identify possible faked data sets. However, the method might be a valid and reliable approach to capture ERS and MPR and to correct for these biases. A limiting factor of the present study is that only laboratory data from a student sample and one specific questionnaire could be used. Thus, replications with real-life data from applicant settings and representative data from large-scale surveys including other measures than the NEO-PI-R are necessary. Moreover, future research should investigate how scores of multi-item scales corrected for response bias in MRM analyses can improve data quality, e.g. by purging irrelevant variance from measures and improving their statistical relationship with important outcome variables.

Nevertheless, the finding that ERS and MPR also exist on the level of the interviewee within a survey bears implications for survey research in general and identification of protocols faked by the interviewer in particular. Methods counting on ERS as being indicative of wrong doing on the side of the person conducting the interview have to take into account that such a response style can and most likely will also occur due to the person being interviewed. Thus, MRM as used here might still be a helpful tool in order to ascertain that a response style is due to the interviewee and not the interviewer. Future research could therefore use data sets including faked protocols and apply MRM to potentially differentiate between response styles on the side of the interviewee and wrong doing by the interviewer.

Summing up, the present study indicates that MRM are not suitable to detect faking. However, identifying ERS and MPR reliably seems feasible and offers the opportunity to correct data for the detrimental effects such response styles might have and to improve the quality of survey data based on multi-point rating scale items.

Bibliography

Amthauer, R., Brocke, B., Liepmann, D., & Beauducel, A.: I-S-T 2000 R (Intelligenz-Struktur-Test 2000 R) [Intelligence-Structure-Test 2000 R], Göttingen, Hogrefe, 2001.

Austin, E. J., Deary, I. J., & Egan, V.: Individual differences in response scale use: Mixed Rasch modelling of responses to NEO-FFI items, Personality And Individual Differences, 40, 1235-1245, 2006.

Berg, I. A., & Collier, J. S.: Personality and group differences in extreme response sets, Educational and Psychological Measurement, 13, 164-169, 1953.

Birkeland, S. A., Manson, T. M., Kisamore, J. L., Brannick, M. T., & Smith, M. A.: A meta-analytic investigation of job applicant faking on personality measures, International Journal of Selection and Assessment, 14, 317-335, 2006.

Bolt, D. M., & Johnson, T. R.: Addressing score bias and differential item functioning due to individual differences in response style, Applied Psychological Measurement, 33, 335-352, 2009.

Bozdogan, H.: Model selection and Akaike's Information Criterion (AIC): The general theory and its analytical extensions, Psychometrika, 52, 345-370, 1987.

Bredl, S., Winker, P., Kötschau, K.: A statistical approach to detect interviewer falsification of survey data, Survey Methodology, 38, 1-10, 2012.

Costa, P. T., & McCrae, R. R.: Revised NEO personality inventory (NEO-PI-R) and NEO Five-Factor inventory (NEO-FFI): Professional Manual.: Odessa, Psychological Assessment Resources, 1992.

Eid, M., & Rauber, M.: Detecting measurement invariance in organizational surveys, European Journal of Psychological Assessment, 16, 20-30, 2000.

Eid, M., & Zickar, M.: Detecting response styles and faking in personality and organizational assessments by mixed Rasch models, In von Davier, M. & Carstensen, C. H. (eds.), Multivariate and mixture distribution Rasch models (255-270), New York, NY, Springer, 2007.

Ellingson, J. E.: People fake only when they need to fake. In Ziegler, M., MacCann, C. & Roberts, R. (eds.), New perspectives on faking in personality assessment (19-33), New York, NY, Oxford University Press, 2011.

Gerber-Braun, B.: The Double Cross. Dissertation, LMU München, Fakultät für Psychologie und Pädagogik, 2010.

Gollwitzer, M., Eid, M., & Jürgensen, R.: Response styles in the assessment of anger expression, Psychological Assessment, 17, 56-69, 2005.

Greenleaf, E. A.: Measuring extreme response style, Public Opinion Quarterly, 56, 328-351, 1992.

Grubb, W. L., & McDaniel, M. A.: The fakability of Bar-On's Emotional Quotient Inventory Short Form: Catch me if you can, Human Performance, 20, 43-59, 2007.

Heggestad, E. D., George, E., & Reeve, C. L.: Transient error in personality scores: Considering honest and faked responses, Personality and Individual Differences, 40, 1201-1211, 2006.

Hinz, A., Brahler, E., Geyer, M., & Korner, A.: Urteilseffekte beim NEO-FFI / Response sets measures with NEO-FFI, Diagnostica, 49, 157-163, 2003.

Kemper, C.: Das Persönlichkeitsmerkmal Angstsensitivität: Taxon oder Dimension? – Eine Analyse mit dem Mischverteilungs-Raschmodell, Kovacs, 2010.

Kruglanski, A. W.: Lay epistemics and human knowledge: Cognitive and motivational bases, Plenum Press, 1989.

Kulas, J. T., & Stachowski, A. A.: Middle category endorsement in odd-numbered Likert response scales: Associated item characteristics, cognitive demands, and preferred meanings, Journal of Research in Personality, 43, 489-493, 2009.

MacCann, C., Ziegler, M., & Roberts, R.: Faking in personality assessment: Reflections and recommendations, In Ziegler, M., MacCann, C. & Roberts, R. (eds.), New perspectives on faking in personality assessment (309-329), New York, Oxford University Press, 2011.

McFarland, L. A., & Ryan, A. M.: Variance in faking across noncognitive measures, Journal of Applied Psychology, 85, 812-821, 2000.

Meisenberg, G., & Williams, A.: Are acquiescent and extreme response styles related to low intelligence and education? Personality and Individual Differences, 44, 1539-1550, 2008.

Meiser, T., & Machunsky, M.: The personal structure of personal need for structure - A mixture-distribution Rasch analysis, European Journal of Psychological Assessment, 24, 27-34, 2008.

Mueller-Hanson, R., Heggestad, E. D., & Thornton, G. C.: Individual differences in impression management: an exploration of the psychological processes underlying faking, Psychology Science, 3, 288-312, 2006.

Naemi, B. D., Beal, D. J., & Payne, S. C.: Personality predictors of extreme response style, Journal of Personality, 77, 261-286, 2009.

Ostendorf, F., & Angleitner, A.: NEO-PI-R. NEO Persönlichkeitsinventar nach Costa und McCrae. Revidierte Fassung. [NEO-PI-R. NEO Personality Inventory], Göttingen: Hogrefe, 2004.

Paulhus, D. L.: Measurement and control of response bias. In Robinson, J.P. & Wrightman, L.S. (eds.), Measures of Personality and Social Psychological Attitudes (17-59), San Diego, CA , Academic Press, 1991.

Paulhus, D. L.: Socially desirable responding: The evolution of a construct. In Braun, H. I., Jackson, D. N. & Wiley, D. E. (eds.), The role of constructs in psychological and educational measurement (49-69), Mahwah, NJ, Lawrence Erlbaum Associates, 2001.

Pauls, C. A., & Crost, N. W.: Cognitive Ability and Self-Reported Efficacy of Self-Presentation Predict Faking on Personality Measures, Journal of Individual Differences, 26, 194-206, 2005.

Read, T. R. C., & Cressie, N.: Goodness-of-Fit Statistics for Discrete Multivariate Data, New York, NY, Springer-Verlag, 1988.

Robie, C., Brown, D. J., & Beaty, J. C.: Do people fake on personality inventories? A verbal protocol analysis, Journal of Business and Psychology, 21, 489-509, 2007.

Rogers, R.: Researching Dissimulation. In Rogers, R. (ed.), Clinical assessment of malingering and deception (2 ed., 309-327), New York, NY, Guilford, 1997.

Rost, J., Carstensen, C. H., & Von Davier, M.: Applying the mixed Rasch model to personality questionnaires. In Rost, J. & Langeheine, R. E. (eds.), Applications of latent trait and latent class models in the social sciences, New York, NY, Waxmann, 1997.

Rost, J., Carstensen, C. H., & Von Davier, M.: Are the Big Five Rasch scalable? A reanalysis of the NEO-FFI norm data, Diagnostica, 45, 119-127, 1999.

Von Davier, M.: WINMIRA 2001, Groningen, The Netherlands, ASC-Assessment Systems Corporation USA and Science Plus Group, 2001.

Zickar, M. J., & Burnfield, J. L.: A Monte Carlo investigation of the polytomous mixed Rasch model, unpublished manuscript, 2003.

Zickar, M. J., Gibby, R. E., & Robie, C.: Uncovering faking samples in applicant, incumbent, and experimental data sets: An application of mixed-model item response theory, Organizational Research Methods, 7, 168-190, 2004.

Ziegler, M.: Situational demand and its impact on construct and criterion validity of a personality questionnaire: State and trait, a couple you just can't study separately! Dissertation, LMU München, Fakultät für Psychologie und Pädagogik, 2007.

Ziegler, M.: Applicant Faking: A Look Into the Black Box, The Industrial and Organizational Psychologist, 49, 29-36, 2011.

Ziegler, M., & Bühner, M.: Modeling Socially Desirable Responding and Its Effects, Educational and Psychological Measurement, 69, 548-565, 2009.

Ziegler, M., Danay, E., Schölmerich, F., & Bühner, M.: Predicting academic success with the Big 5 rated from different points of view: Self-rated, other rated and faked, European Journal of Personality, 24, 341-355, 2010.

Ziegler, M., MacCann, C., & Roberts, R. R.: Faking: Knowns, unknowns, and points of contention. In Ziegler, M., MacCann, C. & Roberts, R. (eds.), New perspectives on faking in personality assessment (309-329), New York, Oxford University Press, 2011.

List of Contributors

Ackermann, Daniela, Survey Design and Methodology, GESIS – Leibniz-Institute for the Social Sciences Mannheim, B2 1, 68159 Mannheim, daniela.ackermann@gesis.org

Blasius, Jörg, Department of Political Science and Sociology, Rheinische Friedrich-Wilhelms-University Bonn, Lennéstrasse 27, 53113 Bonn, jblasius@uni-bonn.de

Bredl, Sebastian, Department of Economics, Justus-Liebig-University Giessen, Licher Strasse 64, 35394 Giessen, sebastian.bredl@wirtschaft.uni-giessen.de

Brüderl, Josef, Department of Sociology, Ludwig-Maximilians-University München, Konradstrasse 6, 80801 München, josef.bruederl@soziologie.uni-muenchen.de

Gwartney, Patricia A., Department of Sociology, 613 PLC, University of Oregon, Eugene, OR 97403, pgwartney@gmail.com

Haan, Marieke, Department of Communication and Information Sciences, University of Groningen, Oude Kijk in 't Jatstraat 26, 9712 EK Groningen, marieke.haan@rug.nl

Hülser, Oliver, GfK SE Marketing Science, Nürnberg, Nordwestring 101, 90149 Nürnberg oliver.huelser@gfk.com

Huiskes, Mike, Department of Communication and Information Sciences, University of Groningen, Oude Kijk in 't Jatstraat 26, 9712 EK Groningen, m.huiskes@rug.nl

Huyer-May, Bernadette, Department of Sociology, Ludwig-Maximilians-University München, Konradstrasse 6, 80801 München, bernadette.huyer-may@soziologie.uni-muenchen.de

Jesske, Birgit, infas – Institute for Applied Social Sciences, Friedrich-Wilhelm-Strasse 18, 53113 Bonn, b.jesske@infas.de

Kemper, Christoph J., Institute for Medical and Pharmaceutical Proficiency Assessment, Große Langgasse 8, 55116 Mainz, contact@christoph-kemper.net

Martin, Silke, Survey Design and Methodology, GESIS – Leibniz-Institute for the Social Sciences Mannheim, B2 1, 68159 Mannheim, silke.martin@gesis.org

Massing, Natascha, Survey Design and Methodology, GESIS – Leibniz-Institute for the Social Sciences Mannheim, B2 1, 68159 Mannheim, natascha.massing@gesis.org

Menold, Natalja, Survey Design and Methodology, GESIS – Leibniz-Institute for the Social Sciences Mannheim, B2 1, 68159 Mannheim, natalja.menold@gesis.org

Ongena, Yfke, Department of Communication and Information Sciences, University of Groningen, Oude Kijk in 't Jatstraat 26, 9712 EK Groningen, y.p.ongena@rug.nl

Opper, Marie K., Center for International Environmental and Development Research, Justus-Liebig-University Giessen, Senckenbergstrasse 3, 35390 Giessen, marie.k.opper @sowi.uni-giessen.de

Porst, Rolf, Survey Design and Methodology, GESIS – Leibniz-Institute for the Social Sciences Mannheim, B2 1, 68159 Mannheim, rolf.porst@gesis.org

Rammstedt, Beatrice, Survey Design and Methodology, GESIS – Leibniz-Institute for the Social Sciences Mannheim, B2 1, 68159 Mannheim, beatrice.rammstedt@gesis.org

Schmiedeberg, Claudia, Center for European Social Research Mannheim (MZES), A5 6, 68159 Mannheim, claudia.schmiedeberg@mzes.uni-mannheim.de

Slavec, Ana, Center for Social Informatics, University of Ljubljana, Kardeljeva ploščad 5, 1001 Ljubljana, ana.slavec@fdv.uni-lj.si

Storfinger, Nina, Center for International Environmental and Development Research, Justus-Liebig-University Giessen, Senckenbergstrasse 3, 35390 Giessen, nina.storfinger @zeu.uni-giessen.de

Thiessen, Victor, Department of Sociology and Social Anthropology, Dalhousie University, 6135 University Ave, Rm 1128 Halifax, thiessen@dal.ca

Vehovar, Vasja, Center for Social Informatics, University of Ljubljana, Kardeljeva ploščad 5, 1001 Ljubljana, vasja@fdvinfo.net

Winker, Peter, Center for International Environmental and Development Research, Justus-Liebig-University Giessen, Senckenbergstrasse 3, 35390 Giessen, peter.winker@wirtschaft.uni-giessen.de

Zabal, Anouk, Survey Design and Methodology, GESIS – Leibniz-Institute for the Social Sciences Mannheim, B2 1, 68159 Mannheim, anouk.zabal@gesis.org

Ziegler, Matthias, Department of Psychology, Humboldt-University Berlin, Unter den Linden 6, 10099 Berlin, matthias.ziegler@psychologie.hu-berlin.de

Schriften zur empirischen Wirtschaftsforschung

Herausgegeben von Peter M. Schulze und Peter Winker

Band 1 Christoph Balz: Multivariate Überprüfung von Hysteresiseffekten. Eine empirische Analyse ausgewählter Arbeitsmärkte. 1999.

Band 2 Daniel Porath: Fiskalische Beurteilung der Staatsverschuldung mit ökonometrischen Methoden. Eine empirische Studie für die Bundesrepublik Deutschland. 1999.

Band 3 Martina Johannsen: Theorie und Empirie von Arbeitsmärkten. Eine ökonometrische Analyse für Rheinland-Pfalz. 2000.

Band 4 Peter M. Schulze: Regionales Wachstum und Strukturwandel. Quantitative Analyse mit Regionaldaten für die Bundesrepublik Deutschland. Unter Mitarbeit von Christoph Balz. 2001.

Band 5 Nora Lauterbach: Tertiarisierung und Informatisierung in Europa. Eine empirische Analyse des Strukturwandels in Deutschland, Frankreich, Italien und Großbritannien. 2004.

Band 6 Robert Skarupke: Renditen von Bildungsinvestitionen. Paneldaten-Schätzungen für die Bundesrepublik Deutschland. 2005.

Band 7 Manfred Scharein: Zur Theorie skalenparametergesplitteter Verteilungen und ihrer Anwendung auf den deutschen Aktienmarkt. 2005.

Band 8 Ke Ma: Quantitative Renditeanalysen am deutschen Aktienmarkt mit Multifaktoren-Modellen. 2005.

Band 9 Jens Ulrich Hanisch: Rounding of Income Data. An Empirical Analysis of the Quality of Income Data with Respect to Rounded Values and Income Brackets with Data from the European Community Household Panel. 2007.

Band 10 Yvonne Lange: Fertilität und Erwerbsbeteiligung von Frauen in Deutschland. Eine empirische Analyse. 2007.

Band 11 Jörg Schmidt: Relative Deprivation, Arbeitszufriedenheit und Betriebswechsel. Eine Analyse auf Basis von Linked Employer-Employee Daten. 2008.

Band 12 Tanja Kasten: Monetäre und nicht-monetäre Effekte von Erwerbsunterbrechungen. Eine mikroökonometrische Analyse auf Basis des SOEP. 2008.

Band 13 Alexander Prinz: Quantitative Analysen zum deutschen und internationalen Luftfrachtmarkt. 2008.

Band 14 Anke Koch: Regionale Verteilung der Beschäftigung in Deutschland. Panel- und Zähldatenmodelle. 2009.

Band 15 Martin Mandler: Geldpolitische Reaktionsfunktionen und makroökonomische Unsicherheit. 2010.

Band 16 Verena Dexheimer: Einreisetourismus in Deutschland. Paneldatenanalysen und SARIMA-Prognosen. 2010.

Band 17 Julia König: Gesundheitsausgaben in Deutschland. Eine Kointegrationsanalyse. 2010.

Band 18 Akwum Onwunta: Contributions to Credit Portfolio Modeling and Optimization. 2011.

Band 19 Vahidin Jeleskovic: Agentenbasierte Modelle für empirische Wechselkurse. Ökonometrische Schätzung und Evaluierung. 2011.

Band 20 Jörg Böttcher: Möglichkeiten einer Projektfinanzierung bei CSP-Vorhaben. 2012.

Band 21 Ivan Savin: Russia´s Comparative Advantages in Foreign Trade. Forecasting Potential Dynamics and Identifying Stimulation Measures. 2013.

Band 22 Peter Winker/Natalja Menold/Rolf Porst (eds.): Interviewers´ Deviations in Surveys. Impact, Reasons, Detection and Prevention. 2013.

www.peterlang.de

www.ingramcontent.com/pod-product-compliance
Ingram Content Group UK Ltd.
Pitfield, Milton Keynes, MK11 3LW, UK
UKHW020258051026
15470UKWH00021B/1366

9 783631 637159